Jerald G. Schutte

Columbia University

Everything You Always Wanted To Know About Elementary Statistics
(but were afraid to ask)

Prentice-Hall, Inc., Englewood Cliffs, New Jersey 07632

Library of Congress Cataloging in Publication Data

SCHUTTE, JERALD G
    Everything you always wanted to know about elementary statistics (but were afraid to ask).

    Includes bibliographies and index.
    1. Statistics. I. Title.
HA29.S425 1977   519.5   76-46638
ISBN 0-13-293506-6

PRENTICE-HALL METHODS OF SOCIAL SCIENCE SERIES

Herbert L. Costner and Neil Smelser, *Editors*

© 1977 by Prentice-Hall, Inc., Englewood Cliffs, New Jersey 07632

Printed in the United States of America

10 9 8 7 6 5 4 3

Prentice-Hall International, Inc., *London*
Prentice-Hall of Australia Pty. Limited, *Sydney*
Prentice-Hall of Canada, Ltd., *Toronto*
Prentice-Hall of India Private Limited, *New Delhi*
Prentice-Hall of Japan, Inc., *Tokyo*
Prentice-Hall of Southeast Asia Pte. Ltd., *Singapore*
Whitehall Books Limited, Wellington, *New Zealand*

Contents

iii

III

Inferences

IV

Nonparametric alternatives

Appendixes

Prologue

As a social science researcher, but more important, a teacher, I suppose I should be horrified by the plague of panic and hostility that infects so many students having to study social science courses dealing with statistics. However, in a very real sense, I believe there are two fundamental reasons for this epidemic.

The first element reflects an increasing alienation from the use of numbers and statistics to describe the human condition. From the moment a baby is fitted with plastic wristband containing a number matching his mother's and identifying parentage, he is packaged, filed, and labeled in a hundred different numerical ways. Despite our impotent protests, society assigns us numerical identity on student cards, library identification, intelligence tests, driver's licenses, social security records, and even graveyard plots. We are further asked, almost daily, to internalize large doses of numerical information from all forms of mass media. We must understand weather reports, ozone counts, television ratings, political polls, and market pricing. We are forced to accept governmental decisions on the basis of statistical averages, not human emotions, and mathematical projections, not individual needs. Even our leisure activities seem bound up in numerical bureaucracy: fishing licenses, camping reservations, and even lotteries to determine who receives season tickets to athletic events. Is it any wonder many people tend to grow up with an increasing distaste for the use of numbers?

Although the reasons for this distaste are many, they seem to be compounded by a second element. This problem reflects several basic obstacles in learning to understand, accept, and use mathematics. The first obstacle is the public school system's apparent inability to pace the introduction of mathematical concepts with a child's developing capacity to comprehend. Students' introduction to the most abstract, and yet the most basic assumption of mathematics, the number system, comes at a time when they are least able to cope with abstractions. When we are taught to count, the use of ice cream sticks and tongue depressors as the unit, tens', and hundreds' place holders more likely conjures up a picture of popsicle orgies or dreaded trips to the doctor's office than integers in the real number system. As the education continues, the notion of pie charts seemingly evokes more pangs of hunger than visions of fractions. These valid but poorly timed techniques are continued over the years, interrupted at regular intervals by assignments not quite comprehended and tests on material not quite mastered. No wonder the unfortunate student despairs of ever understanding mathematics—he has become too bogged down in understanding the preliminary assumptions.

A second obstacle is that mathematics, like many other skills, involves certain essential chores of memorization. Even the advanced mathematician must consciously remember that $7 \times 8 = 56$. It is hardly possible to continue understanding higher-order abstractions unless we have the basic computations of addition, subtraction, multiplication, and division memorized. Yet we ask children to memorize these tables at a point in their life when it seems the attention span is at a minimum. It is small wonder that when later called upon to perform, our thought process creates anxiety over the inability to remember such seemingly simple things.

A third problem is generated by the weakness students often sense in the teacher's own grasp of mathematics. Sketchy explanations, vague instructions to "do the problem the best you can; you'll probably understand it as you do the work" deceives no one. Arbitrary conventions and lack of intuitive reasoning handicap the very authority whose function it is to make all these things clear. The message to the student becomes almost a self-fulfilling prophecy of doubt, rejection, and failure.

The fourth obstacle appears to be one brought about by fear of labeling within the peer group. If confusion and avoidance of math become the norm, words like "egghead," "brain," or "kissy" (to mention the least offensive) may be attached to those who have developed the reasoning abilities to cope with numbers. For females, the emphasis is often on social conformity and may entail the tacit awareness that doing well in math class risks *not* doing well with the men in that class.

For males, there is often a masculine mystique that emphasizes success with women and athletics; academic grinding is to be avoided at all costs.

One major consequence of these obstacles is that many college freshmen, having been caught up in this pattern, tend to draw away from the hard sciences into the social sciences, which have the reputation of being less quantitative. A great many of these students, convinced they lack the ability to deal with numbers, soon discover that practically every social science major requires a class in statistics; the ensuing encounter with the necessary statistics course becomes an almost universally dreaded and demoralizing walk down the path to pessimism and panic. The accompanying symptoms of ulcers, insomnia, and tears are often perpetuated by putting the class off until the senior year, where it becomes mandatory to pass in order to graduate.

This book is written with these students in mind. That is, it is a supplementary text for students in social science statistics classes. It is not meant to replace existing texts on the subject, although in some courses it could be relied upon to provide the main introduction to the field. Moreover, it is not intended to present a detailed mathematical treatment of the subject. Indeed, its purpose is to avoid the tortuous mathematical theorems and assumptions that account for most of the student's hesitations already. However, the book does not try to eliminate the mathematics (as is often done) in treating a statistical application. Rather, it periodically introduces the mathematical logic behind some formulae and derivations. By doing so, in a number of the simpler cases, it is hoped that the reader will generalize the comfort found in understanding these derivations to situations where the formulae appear more complex.

The technique used is a question and answer format. This is not meant to be a programmed approach. Rather, the manuscript actually derives from students over the past several years. These are their questions, and to a large extent, their answers. What I have tried to do is compile them in a coherent manner and provide supplementary information where I felt it was needed. The educational device employed is not the typical social science example used to illustrate certain statistical tools. Instead, I have tried wherever possible to introduce topics with the literary technique of analogy and metaphor. The reasoning is that a statistical concept (like any other concept) is rendered much more intuitive if removed from its world of mathematical symbols and anchored in the perceptual experience of the individual's everyday life. Thus, the student with a mental block for mathematics is more apt to remember the term *mean* by recalling a teeterboard or see-saw at the playground, or the concept of a *distribution* by imagining a roller-coaster, or the rationale for *permutation* by thinking of horse races, than he ever will by drawing on his storehouse of partially remembered mathematical derivations.

The approach does not end here, however. After the concept is defined and an analogy is established, the symbolic formula and a more traditional example are elaborated. Where feasible, the mathematical rationale is explained. Lastly, reference to the ad hoc and historical nature of statistical terms and symbols is made in the Appendixes.

I believe this approach has three benefits. As a supplement, the student will want to refer to it for specific information. The question and answer format, broken down by chapter and following a logical progression of ideas, provides a more readable reference. Secondly, by providing a brief summary of the history of the methods and symbols, some of the mystique is taken from statistical analysis. Finally, by approaching the subject matter in a less complicated manner, the student is likely to internalize statistics with less distaste for the mathematics involved.

Part I serves as an introduction to the subject matter, focusing on the basic vocabulary of statistics (chap. 1) and the various levels of measurement (chap. 2). Part II deals with descriptive statistics. Having made the distinction between categorical and quantitative measurement, I treat them in chapters 3 and 4 respectively with discussions of the tables and graphs for each. Chapter 5 begins with a statement on central tendencies and chapter 6 relates these to measures of dispersion, including the concept of z-scores. Finally, chapters 7 and 8 introduce techniques useful for describing two quantitative variables, elaborating the ideas behind regression and correlation respectively.

Part III provides an introduction to hypothesis testing. In chapters 9 and 10 probability and probability distributions, such as the binomial, are related to the normal curve. Chapter 11 enumerates some of the ground rules for testing hypotheses when using sampling distributions. Chapters 12 to 14 treat specific cases of these tests. Chapter 12 illustrates the single sample case of testing and estimation. Chapter 13 explains the sampling distribution of the difference of means and introduces the F distribution. Chapter 14 extends the use of the F distribution and presents an example of simple one-way analysis of variance.

Part IV discusses special nonparametric techniques. In chapter 15 nonparametric tests of hypotheses are illustrated, and in chapter 16 traditional nonparametric relational measures are discussed. Finally, four appendixes are included: (I) historical notes, (II) a discussion of relevant statistical symbols, (III) a review of forgotten algebraic operations and tips on use of the slide rule and calculators, and (IV) a set of tables of statistical inference.

The book can be utilized in several different ways. As an introduction, students may wish to read it first to introduce themselves to the subject matter in a main text. As a reference, the book can serve as further explanation for individual questions. As a supplement, students may utilize specific sections, such as the historical notes, the appendixes on algebra, slide rule, and statistical symbols, or the selected reference sections at the end of each chapter.

I have undoubtedly left out material that some will deem important. I may have included information that some will think is extraneous. However, in an effort to coordinate this subject matter with existing texts in the social sciences, I have tried to promote two ideas: never present a statistical technique until it is intuitively expressed; and always anchor this intuition in an analogy representing the everyday experiences of the student.

Acknowledgments

I wish to acknowledge the critical support of a number of people: Dr. Herbert Costner, for his copious reviews of several versions of this manuscript; Ed Stanford and his staff at Prentice-Hall, for their undying patience and renewed dedication to the concepts underlying this effort; James Keeshan, for his self-sacrificing service in producing the cartoon illustrations; John Light, for his continued support and feedback during the last stages of writing; and finally, my family, friends, and colleagues, without whose continued devotion and understanding I would have never mustered the motivation to put my ideas on paper.

JERALD SCHUTTE
Los Angeles, California

Introduction

I

Numbers, symbols, and words

A STUDENT'S LAMENT

1

I don't understand it! I don't understand it!

Why does a person need to study statistics?

Like it or not, you have been exposed to certain forms of statistics the better part of your life. Most of what we call current events are made known to us via such reports as opinion polls, stock averages, unemployment rates, and Gross National Products. Sports and weather reports have familiarized us with concepts such as averages, percentages, and probabilities. And advertising has sensitized us to the differences between groups with fewer cavities, shinier hair, and softer hands.

Yet in each one of these political, social, and economic contexts, the availability of masses of information suggests a great deal of confusion and misuse. Therefore, it is essential for the serious student interested in contemporary society to gain some understanding of the concepts involved in the study of statistics, both to guard against inaccurate information and to make intelligent decisions.

Do I have to be good in math to understand statistics?

The language of statistics is powerful, yet the mathematical operations are as simple as those encountered in the fourth grade. At no time are we asked to deal with more than four basic operations: addition, subtraction, multiplication, and division. At first glance, these may appear to be combined in strange ways (e.g., a special case of dividing is finding the square root; a special case of multiplying is finding the power of a value). They may even be used to produce strange-sounding terms (e.g., standard error of the mean). But you will never be asked to compute anything that does not boil down to the four basic operations.

But I can't seem to work with numbers!

Many students take this attitude in social science statistics classes. The pattern appears to be the following: the first day of class there is a feeling of guarded optimism, "I can do it, I can do it." Tinges of panic set in after a lecture or two, "I must do it, I must do it." Then, the first homework assignment is encountered and the feeling is, "I don't understand it, I don't understand it." Finally, a missed lecture or a disappointing talk with the professor, and the student has confirmed what he suspected in the beginning, "I'm no good with numbers, I knew it; I'm just not good with mathematics at all." This problem is often intensified when the dreaded statistics requirement is ignored, only to be taken in the senior year. By this time the lag between the last high school math class and this sadistic requisite may have been four to six years.

The experience is particularly unfortunate because the problem is not one of having to feel comfortable with numbers specifically, or statistics in general. Rather it is to feel comfortable with the idea of using numbers and symbols, much the same as we do words, in creating a statistical language which describes and predicts events in the "real" world.

Is statistics really a language or just masses of information?

Don't confuse *what* is being measured with *how* it is measured. Information such as unemployment or birth rates may communicate more or less depending upon the context and how it is put to use. This has very little to do with the means by which it is measured. The study of statistics provides us with a set of symbols and operations which become the means to accomplish this measurement. Just as words are labels which represent certain tangible or abstract concepts, statistical symbols and operations are also labels which can be used to eliminate some of the problems brought about by the ambiguity of using words to describe information and infer properties from it. When we combine numbers and symbols in certain specified ways, deriving more powerful concepts (e.g., mean and variance) and represent them with various Greek and English letters, we have an impressive language. And that is precisely what statistics is, a language. In everyday English, we describe situations by pointing out similarities and differences among events; statistics accomplishes the same goal. The beauty of a statistical vocabulary is that we can do this in a much more precise, less inconsistent manner than is possible through the use of everyday language.

As mentioned earlier, there are four basic mathematical symbols in the vocabulary of statistics ($+$, $-$, $\times$, and $\div$), one corresponding to each operation. However, to make a statement in the vocabulary we introduce a fifth symbol which acts as the verb in our language; the sign is equality ($=$). We can now combine symbols and numbers to make statements: $4 + 5 = 9$. This is a statement about the sum of 4 and 5. However, we could make this even more abstract by letting other symbols stand for the numbers: $x + y = z$. This is a general case in which we can substitute the particular values we happen to have. In the vocabulary of statistics, we represent combinations of general statements (composed of symbols and operations) by other symbols, e.g., $\bar{X} = \Sigma X_i / N$ and $s^2 = \Sigma (X_i - \bar{X})^2 / N - 1$. In this way we build each successive concept on those preceding it. Each term is given a name and it is these names which comprise the vocabulary of statistics.

However, just as we have basic symbols corresponding to the primary operations, so we have several basic definitions which define other aspects of the vocabulary of statistics. As in all logical systems, these definitions are givens and we must learn their meaning to use the language. We begin with the most general word statistics.

What does the word statistics *mean?*

Statistics (plural) is a branch of applied mathematics specializing in procedures for describing and reasoning from observations. It is generally acknowledged to be divided into two areas: descriptive and inferential statistics. The function of **descriptive statistics** is to describe observations collected in populations and samples. The function of **inferential statistics** is to infer something about populations, given descriptions of a sample.

How do I know whether I'm dealing with a population or a sample?

Any collection of objects, events, people, etc., which is defined because of some uniqueness, is called a **population**. The number of whales in the world can be a population as can the number of students in the class. The number of elements in a compound as well as the number of people at your next party qualify as populations. In short, a population is what you choose it to be. It is the unit whose characteristics you care to describe by observing them, directly or indirectly.

Unfortunately, some populations can be quite unwieldy. It would be difficult, at best, to measure something about all the men in the world—height, for example. Not only would it be expensive, but I dare say you may get lost trying to find all of them. Luckily, however, we can solve this problem by looking at a sample of them. A **sample** is a subset of a population. By looking at this subset and describing it, we can infer something about the larger population from which it came.

It is easy to see that at one time the students in your class may qualify as a sample, while at another they could be a population. The difference is whether you care to generalize the results to a larger group (as with a sample) or are content to describe only those you have selected (a population).

Why are different symbols used for populations and samples and what is the distinction?

A symbol used to indicate the measurement of a variable in a population is called a **parameter**. A symbol used to represent a given measure in a sample is called a **statistic** (singular). Greek letters are generally used for population parameters and English letters are used for sample statistics. In describing variables in a population and a sample we use the same operations; the only difference is the symbol. Actually, even the use of symbols is quite arbitrary and stems from a strong tradition in mathematics in which Greek letters are extensively used. In descriptive statistics they often appear interchangeably. But their major function is to allow us to make the distinction between merely describing variables and attributes or using those descriptions to infer something about a larger population.

Are variables and attributes different names for the same concept?

Any particular trait which can take on a range of values in a population or sample is called a **variable**. The height of males in the world is a trait which takes on a range of values. Some men are very short and some are very tall. The number system we impose is feet and inches, or in some countries, centimeters. An **attribute**, on the other hand, represents a variable whose variation is among different categories rather than on some quantitative continuum. Religion as an attribute can be conceived as Protestant *or* Catholic *or* Jewish, etc.

Because there are a number of traits that can vary in a collection, we have a number of variables. Think of how many characteristics differ among you and your friends: age, weight, height, coloring of hair and eyes, the type of car you drive, grade point average, and a hundred others. **Measurements** of selected variables for a given set of cases are called **data**. Therefore, when we say that we have data on a particular population or sample, we mean that we have measured certain variables in that collection.

What is the best way to study statistics?

There is a basic process for understanding any new subject matter. However, it is particularly useful in understanding statistics. First, you must be able to recognize and identify the concepts. This means you need to be familiar with the words and symbols which define each concept. Begin by skimming appendix II to familiarize yourself with the notation. Become *aware* of it without memorizing. Second, understand the properties and relations of the new concepts. We stated earlier that each term is built on the combination of previously used symbols and words. Recognize their relationship and understand the operations of combining symbols. Go to appendix III, on algebra review, to improve these skills. Third, seek out an analogy in your everyday experience with which to associate new words in the vocabulary of statistics. Commit this association to memory. Fourth, find as many numerical examples of the use of a particular statistical term as possible and work them out. There is no substitute for practice to build a comfortable habit. Examples can be found in any of the references listed at the end of each chapter. Finally, review the four steps several times for each new concept. Terms will rarely make sense the first go-around. However, the more this process is carried out, the easier it becomes. Combining the use of intuitive experience and the knowledge gained from repetition of examples should provide a comfortable beginning to a fascinating subject.

Where do I begin?

We start by explaining four different ways to measure variables. Each successive chapter then begins to explore the statistical concepts used with each type of measurement in describing variables, assessing relationships, and confirming differences.

How Much Do You Remember?

New Words	New Symbols
Statistics	$+$
Descriptive	$-$
Inferential	$\times$
Population	$\div$
Sample	$=$
Parameter	
Statistic	
Variable	
Attribute	
Measurement	
Data	

Did You Ever Wonder?

1. Why is mathematics introduced in elementary school during the first grade?
2. Why doesn't all verbal behavior consist of mathematical symbols?
3. Why is statistics often taught in disciplines other than mathematics?

Want to Know More?

1. Butsch, R. L. *How to Read Statistics.* Milwaukee: Bruce Publishing Co., 1946.
2. Huff, Darrell. *How to Lie with Statistics.* New York: W. W. Norton & Company, Inc., 1954.
3. Kendal, M.G., and W.R. Buckland. *A Dictionary of Statistical Terms.* 2nd ed. London: Oliver & Boyd, 1960.
4. Pearson, Karl. *The Grammar of Science.* 3rd ed. New York: Meridian Books, 1957.

The nature of data

A FLIGHT PLAN
FOR STUDYING STATISTICS

2

What does it mean to say that something is measured?

To measure is to use a number system in describing variables. It is a way of counting or accounting for the things we observe. In social statistics we use number systems for essentially three purposes: (1) to classify things, (2) to order things, and (3) to quantify things. Since "data" are measures on such variables, we often refer to them as (1) nominal data, (2) ordinal data, and (3) interval or ratio data, respectively.

Are nominal data "measured"?

Most people associate number systems, or what mathematicians like to call the real number system $(0, +1, -1, +2, -2,$ etc., and all fractions in between), with our need to count things: how many cars? how much money? how many babies? The truth is that we use numbers for a lot more than counting. Often we need to account for or classify things and do so by giving them a number. We keep track of television programs by knowing the number of the channel on which they are broadcast. We keep track of cars by assigning each a license plate number. We keep track of communications by numbering our telephones. We account for ourselves by accepting a social security number.

In short, *whenever we assign numbers to a set of categories without reference to direction or magnitude of difference among the alternatives, we are using numbers to measure a* **nominal**-*level variable.* The word nominal comes from the same root as name, and that is what we are doing. We are naming the objects in our collection. We are not implying anything about whether one alternative is more or less than another. Sex, religion, race, political party, and place of birth are all examples of nominal-level variables (attributes).

We account for ourselves by accepting a social security number

Are all attributes classified with numbers?

No. Your name is an example of using letters to classify, distinguishing you from other people. Colors are also used. For example, two competing athletic teams each wear a different color, though the number of people on any given team is usually so large that to use colors to distinguish between each teammate would confuse referees and fans. We really could use any of a variety of systems to classify attributes, but numbers provide a simpler yet more effective way, especially if the set of alternatives is quite large.

When numbers are applied to classifications in statistics,
what do they mean mathematically?

The only mathematical operation that nominal-level data permit is equality ($=$) and inequality ($\neq$). Since we are not measuring the order or the magnitude of the order of categories, numbers are merely used to identify the classification in which an object belongs. Because we are using numbers to qualify the categories, we often refer to these as qualitative data.

Is simple classification like high, medium, *and* low
a measurement?

Yes, but this involves the second purpose of measuring in statistics. *Whenever we assign numbers to order or rank classifications without reference to the magnitude of the difference between ranks, we are using that number system to measure variables as* **ordinal**-*level data.* When we pick the five place winners of a beauty contest or award medals at the Olympics, we are not implying anything about how *much* better one is than the next, only that that person *is* better. If a driver finishes first at the Indy 500, it doesn't matter whether he finished by one car length or five laps! He still wins the same amount of money. Secretariat won the triple crown of horse racing by 31 lengths, but still would have received the prize if victory had been by a nose. Examples of ordinal variables are grades, social class, and prestige.

What mathematical operations do ordinal-level data permit?

Since we are also naming when we rank or order a variable, this level of data allows the operation of equality ($=$) and inequality ($\neq$). But it also refers to the direction of inequality; that is, greater than ($>$), or less than ($<$). First place is greater than second place, which in turn is greater than third place (i.e., lst $>$ 2nd $>$ 3rd). Likewise a general has more prestige than a captain, who in turn has more prestige than a lieutenant (general $>$ captain $>$ lieutenant). This implies that ordered classes are transitive (i.e., if 1st $>$ 2nd, and 2nd $>$ 3rd, then 1st $>$ 3rd). But since we have not made any assumptions about the magnitude of the differences between ranks we still can't use the arithmetic operations ($+$, $-$, $\times$, and $\div$).

How are numbers used to describe quantities?

This is the third function of number systems in statistics. *Whenever we assign numbers to indicate order and assume that the distance between any two consecutive numbers is the same as the distance between any other two consecutive numbers, we are using these numbers to measure variables as* **interval***-level data.* It is a standardized scale of measurement in the sense that within a given context, people agree on the unit of measure. Everybody agrees that when measuring distance, for example, a foot consists of 12 inches, not five toes. Thus, a foot remains 12 inches no matter how many times a ruler is used to measure. Because we are assuming something about the magnitude or quantity of the scale, we call such measurements quantitative data. That is, we are making the assumption that the differences between consecutive positions are equal. We measure the size of families by how many babies they have. We measure income by how many dollars one makes.

Is there a difference between counting and measuring?

We make the distinction between discrete and continuous data. This could be conceived as the difference between counting and measuring. The difference is that **discrete** *data are measured in units which, by definition, cannot be subdivided any further.* There is no such thing as any one family having 2.23 children. Either a family has 2 children or they have 3, but they cannot have anything in between because our measuring unit has no meaning beyond whole numbers.

On the other hand, *units that can take on an unlimited or potentially unlimited number of values are considered* **continuous**. For example, age is usually reported in years, yet if we wanted to, we could break it up into months, days, or even hours. We measure weight in pounds, but could report it to any number of decimal places.

You could argue that the difference is really a matter of the refinement of the measuring instrument. That is, our ability to discriminate is related to how fine our measurement instrument is. Yet some concepts leave no room to refine the unit of measurement: babies in the family, marbles in a bag, cars in the driveway, pennies in the piggybank. The point to remember is that on an interval scale, whatever the degree of accuracy, each unit is the same size.

Are interval and ratio measurement the same?

Both interval and ratio levels are quantitative measurement. The difference between an interval and a ratio scale is that *the interval scale has an arbitrary zero point while the* **ratio** *scale has an absolute zero point.* This is important because an absolute zero point allows us to use the operation of multiplication and division in comparing two different measurements.

If you wanted to measure your height and that of a friend by standing next to a table and measuring each of your heights from the table to the top of the head, you might argue that you are twice as tall as your friend. You would certainly be using interval-level measurement (e.g., inches) but you could hardly say that the tabletop was an absolute zero point. Rather, the floor is the point at which there is no height. Thus, if you used the floor as the starting point and measured both the heights, you might end up saying that you are only one-tenth taller than your friend.

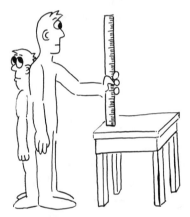

Measuring each of your heights from the table, you might
be tempted to say you are twice as tall as your friend

Taken another way, if you measure temperature on a centigrade scale, zero does not represent the absence of any heat; rather, it is arbitrarily chosen to correspond to the freezing point of water. The absence of all heat is −273° as measured on the Kelvin scale. The centigrade scale is interval-level measurement. The Kelvin scale is ratio measurement. Therefore, it would be misleading if you said that 40° centigrade is twice as hot as 20° centigrade.

In social research, aren't most zero points arbitrary?

The social sciences are a relatively young set of disciplines and the major problems we have are twofold: First, we have difficulty gaining consensus about the definitions of the concepts with which we are involved. Asked to define "love," you may cite one definition while another person may give quite a different answer. Much the same dilemma exists in trying to define such concepts as "prejudice," "occupational prestige," "mobility," or "sentiment." What we do is to rely on operations for definition. An **operational definition** *is a process by which we define concepts on the basis of the operations used to measure them.* For example, I.Q. is defined as a score on a scale designed to tap certain quantitative and verbal skills.

The second problem we have is in developing the appropriate instruments to measure these concepts. Although these are methodological problems, it is important to understand their relation to statistics. In a very real sense, many social researchers have built their reputations by developing ordinal-level instruments to measure nominal-level concepts, and interval-level instruments to measure ordinal-level concepts. Thus, whereas occupation is considered a nominal-level concept, we now measure the prestige of occupations at least ordinally and in some cases as interval measurement. Strength of an attitude is at best ordinal, but certain techniques have been developed to measure them as interval-level data. Even the familiar letter grades (A, B, C, D, F) have had numbers assigned to the letters, and these are treated as interval-level data in computing a grade point average (GPA).

Data like age, income, and education can be conceived to have a zero point. However, because many descriptions do not require it, we often overlook the distinction between interval and ratio in favor of the term quantitative measurement. In computing most social statistics, therefore, the question of an arbitrary or absolute zero point becomes inconsequential.

Do interval- and ratio-level measurement differ
in the mathematical operations permitted?

Just as ordinal scales include the mathematical properties of nominal scales, so interval scales include both the operations of equality (=) and direction of inequality (> <). However, interval scales also permit the operations of addition (+) and subtraction (−). Ratio measurement permits these, but also permits the operations of multiplication (×) and division (÷). Daily temperature as interval-level measurement can tell me how much cooler it is today than it was yesterday, but it cannot tell me that it was "twice" as hot yesterday as last night.

Isn't it confusing to know whether measures are nominal,
ordinal, or interval?

It can be, but as long as you keep three facts straight, you shouldn't have too much trouble. First, if you are merely trying to name, classify, or distinguish alternatives, use nominal-level measurement. Second, if the nature of the data implies magnitude (one alternative has more or less than the next), use ordinal-level measurement. Third, if that order implies equal distances between consecutive units, use interval-level measurement.

How does knowledge of different levels of data contribute to understanding statistics?

Just as each level of measurement implies certain mathematical operations, so these operations for a given level can be combined allowing us to describe and infer things about these particular measurements. In fact, the first question you ask should always be: "What level of data am I dealing with?" Violation of assumptions about which level is used leads to ambiguous conclusions. Hence, it is important to know what level of measurement we are assuming and ascertain what statistics are permissible for describing and inferring from that level. These are the topics to which we now turn.

How Much Do You Remember?

New Words	New Symbols
Nominal	$>$
Ordinal	$<$
Interval	$\neq$
Discrete	
Continuous	
Ratio	
Operational definitions	

Did You Ever Wonder?

1. Why are numbers used to designate athletes on the football field but not for athletes on the tennis court? To designate days in the month but not months of the year?
2. Can data already collected be considered continuous or are they always discrete?
3. What is the difference between numerals and numbers in measurement?

Want to Know More?

1. BLALOCK, H. M., and A. B. BLALOCK *Methodology in Social Research.* New York: McGraw-Hill Book Company, 1968. Chapter 1.

2. CICOURCEL, A. *Method and Measurement in Sociology.* Toronto: The Free Press, 1964. Chapter 1.

3. RUNYON, R. P. and A. HABER. *Fundamentals of Behavioral Statistics.* 2nd ed. Reading, Mass.: Addison-Wesley Publishing Company, 1971. Chapter 2.

4. TORGERSON, W. S. *Theory and Methods of Scaling.* New York: John Wiley & Sons, 1958.

Descriptions

II

Categories
and orders

THE INPUT FOR TABLES
AND GRAPHS

3

How are nominal and ordinal measurements summarized?

The fundamental purpose of describing data is to provide a way of simplifying the information. In nominal-level measurement, we accomplish this by computing frequencies, proportions, percentages, ratios, and rates. In ordinal-level measurement, we can use the additional concepts of cumulative frequency and cumulative percentage. Each of these can be visually represented by tables and graphs.

What does a frequency measure?

Given a set of N numerals (N being the total number of observations) which distinguish items in a collection, we have an **array**; e.g., 1, 2, 1, 1, 3, 2, 3, 3, 2, 1, 1, 3, 3, 2, 2, 2. To summarize this array, we may want to rearrange the numerals so that numbers of the same kind are together: 1, 1, 1, 1, 1, 2, 2, 2, 2, 2, 2, 3, 3, 3, 3, 3. This is called an **ordered array**. Although this helps us to organize the information, it is still in "raw data" form. In nominal-level measurement, we can summarize these "raw data" by reporting how many of each subset or category we have in our data collection; e.g., the number of observations in category 2 is 6. When we organize data in this way, we are indicating frequency. *A **frequency** is the number of observations in any given category of the collection.* It is symbolized by the first letter of the word, a lowercase f_i; where the subscript i indicates any given category. Hence, in the array above, when $i = 2$, $f_i = 6$.

Consider the number of ways you informally keep track of things with the use of frequencies—the number of books you own, pants you wear, telephone calls you make, dates you have, and so on. In each case we use frequency to describe the number of events occurring in each category.

How does the use of frequency help "summarize" data?

First, it allows us to say something about the data without having to enumerate all the items. But more important, it gives us the opportunity to summarize the subsets in terms of the number of observations representing each. We are reporting the frequencies of each category in a set of data. No doubt you have all come in contact with at least a crude form of frequency in your grammar school classroom elections. The teacher would put nominees on the board and keep track with hash marks after each name to indicate the frequency distribution of votes. By comparing the frequencies in the distribution he determined the winner.

We are reporting the frequencies of each category in a set of data

Is frequency the same concept as counting?

It is, in the sense that we are counting up the number of items within each category, not the values of the items (there are no values in nominal-level data). The subtotal number of observations in category 1 is equal to the frequency in category 1. Likewise,

$$\sum_{i=1}^{n} f_i = N \quad \text{or} \quad f_1 + f_2 + \dots + f_n = N$$

That is, the frequency in category 1 plus the frequency in category 2, etc., up to the frequency in the nth category (n being however many categories there are) is equal to the total number of observations N.

In statistics, we use the Greek letter sigma (Σ) to indicate summation. Actually, as late as the 1930's, many used the first letter of the word summation, a capital S. To avoid confusion with other statistical symbols, however, and to be consistent with conventional mathematics, S was slowly replaced by Σ. (See appendix II for more on summations.)

Can frequencies be compared directly?

Yes, but this can be quite misleading. Suppose I want to compare the number of males in two different classes. I might conclude that one class has a higher frequency of males (e.g., 10 in one, 20 in the other), yet proportionately it may be lower (e.g., $^{10}/_{12}$ compared with $^{20}/_{40}$.) When frequencies are expressed as a fraction of the total N, we are reporting the proportion of observations in any given subset. *A* **proportion** *is the frequency of a subset of observations divided by the total number of observations.* Symbolically, $f_i / N =$ the proportion of observations in subset i. Think of a proportion as a piece of pie. Any one piece will be less than or equal to the whole, and all the pieces must equal the entire pie. Likewise, any one proportion is less than (or equal to) 1, and the sum of the proportions must always equal 1. Quite often these proportions are expressed in decimal fractions. That is, if half of the people in your class are males, the proportion is .50.

Can proportions be expressed as percentages?

Yes. *To indicate* **percentage**, *we multiply the decimal fraction by 100.* $^{1}/_{2}$ or .5 $\times$ 100 = 50%. To go from percentages to decimals, we divide by 100. Thus, 30 percent of 100 is .3 or $^{3}/_{10}$. As we have just mentioned, all the pieces must add up to the whole pie, or 100 percent.

This means that no categories or subsets of data may overlap (i.e., they must be mutually exclusive) and they must exhaust all possibilities. In the classroom example, all the people are either boys or girls; this exhausts the possible subsets or categories, and they are mutually exclusive.

What is the difference between a proportion and a ratio?

We said a proportion is any subset divided by the total. In general, a ratio is one number divided by another. However, more often with categorical data, a **ratio** *is any one proportion divided by any other proportion.* One of the best known ratios is used by sociologists and demographers to compare the proportion of men and women in a population. Dividing one proportion into the other, the totals cancel and the result is the frequency of men divided by the frequency of women, i.e., $\dfrac{f_1}{N} \div \dfrac{f_2}{N} = \dfrac{f_1}{\cancel{N}} \times \dfrac{\cancel{N}}{f_2} = \dfrac{f_1}{f_2}$. Expressing this ratio in terms of 100's (i.e., multiplying by 100 to get rid of decimals), we can characterize the sex distribution of a population. A *sex ratio* of 102 means that for every 100 women, there are 102 men in the population. Notice that unlike a proportion, a ratio may be more than 1.

By employing the use of ratios, we have a way of comparing quantities in our collection. In other words we can compare the relative frequency of different subsets of our collection.

Is a rate the same as a proportion?

Similar, yes. The difference is that typically *a* **rate** *divides the frequency of observations by the total and expresses that proportion as a multiple of some standard number.* For example, we may express the number of deaths in a population per each 1000 people. Hence, in any given year we divide the number of people who died by the total population and multiply that proportion by 1000. If in a population of 10,000 200 people died in a year, we would divide 200 by 10,000 and multiply that proportion by 1000. Thus, deaths divided by people times 1000 equals the death rate for a given year expressed per 1000 people, i.e., $200/10{,}000 \times 1000 = 20$.

Examples of rates include birth rates, marriage rates, divorce rates, accident rates, etc., each calculated in reference to a specific base number (e.g., per 1000). One example that helps to illustrate the point is the crime rate. Suppose it were reported that New York city had 900 serious crimes during a given year, and Los Angeles recorded only 600. We might be tempted to say that New York has a more serious crime problem than Los Angeles. However, when we "standardize" by some predefined unit of the population, such as crime per 100,000 people, we find (assuming 9 million people in New York and 5 million people in Los Angeles) that the crime rate for New York is $900/9,000,000 \times 100,000 = 10$, which is smaller than that for Los Angeles at $600/5,000,000 \times 100,000 = 12$.

Are nominal descriptions used for ordinal data as well?

Since numerals used to indicate order include all the mathematical operations used when indicating nominal measurement, we can report all the nominal descriptive techniques for ordinal-level data too. In addition, we can use information about the order of the categories to **cumulate** the frequencies of consecutive classifications. To learn how many people won medals in the Olympics (i.e. placed at least third), we would need to cumulate or add the frequency of people who finished first, second, and third in each event. Likewise, we can cumulate proportions and percentages to determine such things as how many people are passing the class (i.e., getting at least a D or better).

Can qualitative data be summarized in tables?

Academic publications tend to organize descriptions of data into self-contained summaries, and tables have become a practical device for such descriptive accounts. Most of the computer programs in use today (e.g., BMD, SPSS, OSIRIS, and Datatext), plus a wide variety of journals, use somewhat the same format. As table 3-1 indicates, we begin with a listing of the categories down the left-hand column:

TABLE 3-1 Table of Frequencies and Percentages for Student's Marital Status

Value Label	Variable: Marital Status		
	Value	Absolute Frequency	Relative Frequency (percent)
Single	1.00	22	48.9
Married	2.00	18	40.0
Widowed	3.00	2	4.4
Divorced	4.00	3	6.7
	Total	45	100.0

Frequencies and percentages are enumerated in successive columns. We can also include columns for cumulative frequency and/or cumulative percentage. Always, we indicate the total at the bottom, as in table 3-2.

TABLE 3-2 Table of Frequencies and Percentages for Students' Educational Level

| | | | Variable: Year | |
Value Label	Value	Absolute Frequency	Relative Frequency (percent)	Cumulative Frequency (percent)
Freshman	1.00	51	44.7	44.7
Sophomore	2.00	23	20.2	64.9
Junior	3.00	14	12.3	77.2
Senior	4.00	24	21.0	98.2
Graduate	5.00	2	1.8	100.0
	Total	114	100.0	100.0

Can tables be used to summarize more than one variable?

Yes. If two different attributes are being described in a collection, we can present both of them in the same table. Such an arrangement is called a **cross-classification table**. Generally, categories of one variable are listed across the top. Categories of the other variable are listed down the side. Each entry in the table consists of the number of objects in the collection which are contained simultaneously in the row and column category, as in table 3-3.

TABLE 3-3 Cross-classification Table of Frequency for Students' Marital Status by Educational Level

| | Marital Status | | | | |
Year	Single	Married	Widowed	Divorced	Total
Freshman	44	1	0	1	46
Sophomore	22	3	0	1	26
Junior	17	5	1	2	25
Senior	15	7	2	1	25
Graduate	5	14	2	3	24
Total	103	30	5	8	146

In which direction is a cross-classification table percentaged?

It depends on which variable is of interest. A general rule of thumb is to percentage down (i.e., divide the cell frequencies by the column totals and multiply by 100) and compare percentages across, if you are interested in comparing categories of the column variable (table 3-4). Percentage across and compare down if you are interested in comparing categories of the row variable (table 3-5). In

TABLE 3-4 Sex by Political Affiliation

Pol Party	Male	Female	
Repub	N = 12 (50%)	N = 10 (36%)	
Dem	N = 12 (50%)	N = 18 (64%)	
	N = 24 100%	N = 28 100%	N = 52

Percentage ↓ (on left side, pointing down)

TABLE 3-5 Sex by Political Affiliation

Pol Party	Male	Female	
Repub	N = 12 (54%)	N = 10 (46%)	N = 22 100%
Dem.	N = 12 (40%)	N = 18 (60%)	N = 30 100%
Percentage →			N = 52

the first case we are interested in seeing whether there are any differences between males and females (column variable) in terms of party membership; thus we see that 50 percent of the males compared with 36 percent of the females are Republicans. In the second case, we are interested in differences between Republicans and Democrats (row variable) in sex distribution; thus, we see that 54 percent of the Republicans as opposed to 40 percent of the Democrats are males. Later on we will show that differences such as these contribute to the assessment of relationships between two or more variables.

Do graphs also summarize qualitative data?

Yes. Besides presenting qualitative data in tables, which is by far the most common in social science research, we could report the data in the form of pictures or graphs. A **graph** *is a visual representation of a table.* The most common form of graph for qualitative data is the circle or pie graph. We use this whenever we want to represent proportions or percentages (fig. 3-1).

Another kind of visual representation consists of a bar graph, so-called because bars are used to represent the frequency or percentages

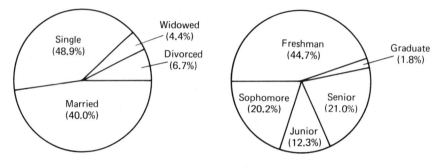

FIGURE 3-1   Percentage of students by marital and educational status

of each category. Most computer programs will put the categories on
the left and the frequency across the bottom as in figure 3-2. However,
when charted by hand, we often put the frequency on the vertical axis
and the categories on the horizontal axis, as in figure 3-3. Notice that
the bars are unconnected across the bottom. This is always done because
we are not implying a continuum or quantitative scale; we imply only
discrete categories which may or may not be ordered.

Often in the mass media and popular literature a representation
known as a pictograph is used in place of a bar graph. Essentially *a*
pictograph *replaces the bars with figures whose heights represent the
frequency of the categories.* The problem here, of course, is that as
the height of the figure is increased, so is the relative width. Thus,
a figure that represents twice the frequency of another may be proportionately four times as large—an obviously misleading presentation. For
such reasons, this technique is rarely used in social science.

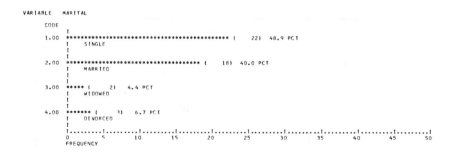

FIGURE 3-2   Frequency by marital status

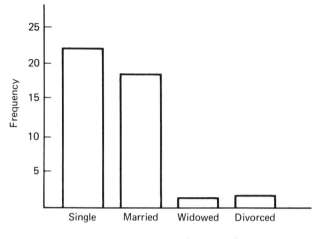

FIGURE 3-3   Frequency by marital status

How do we know which kind of summary to use?

Ultimately, it depends on the circumstance. Graphs are most often used when data are being presented to give a quick visual summary of the proportions or percentages. Tables are most often used when a more precise and detailed summary is desired and where one can spend the time to consider the frequency. In addition, the most important use of a table is when we are presenting data on two or more variables. This would be very difficult to do in graph form for qualitative data.

Do we use the same techniques of table and graph for quantitative data?

Yes. But with an equal-interval scale, we can elaborate our summary. In the next chapter, we use information about the intervals and zero points to aid us in computing percentage tables and constructing bar graphs.

How Much Do You Remember?

New Words	New Symbols
Array	N
Ordered array	n
Frequency	f
Proportion	i
Percentage	$\%$
Ratio	Σ
Rate	
Cumulative frequency	
Cross-classification table	
Graph	
Pictograph	

Did You Ever Wonder?

1. Why, when enumerating a population, do we use a ratio to describe sex composition, but a rate to describe crimes?
2. Why are government budgets most often presented as pie graphs while the stock market uses bar graphs?
3. Is a cross-classification table (i.e., two-way frequency table) more appropriate for nominal than quantitative data?

Want to Know More?

1. ANDERSON, T. R. and M. ZELDITCH. *A Basic Course in Statistics,* 3rd ed. New York: Holt, Rinehart and Winston, Inc., 1975. Chapters 2, 6–7.
2. KLECKA, W. R., N. H. NIE, and C. H. HULL. *SPSS Primer.* New York: McGraw-Hill Book Company, 1975. Chapters 8–9.
3. WEINBERG, G. H., and J. A. SCHUMAKER. *Statistics: An Intuitive Approach.* Belmont, Calif.: Wadsworth Publishing Company, 1962.
4. ZEISEL, HANS. *Say It with Figures.* 5th ed. New York: Harper & Row Publishers, 1968.

Quantitative data

THE GROUPING AND GRAPHING OF NUMBERS

4

Can a table summarize quantitative data?

Since the mathematical operations used with quantitative data include all those used with qualitative data, we can utilize the tools mentioned in the last chapter. However, with the additional assumption of an equal-intervals scale, we can group data, show bar graphs as continuous measures (and therefore employ the frequency polygon), introduce the cumulative-frequency polygon, show trends over time, and weight graphs for disproportionate categories.

Why bother to group data?

We learned in the last chapter that one of the basic ways to summarize data is to list the categories and indicate the frequency distribution of occurrence. In the case of interval-level data, we can do the same thing, except that the categories are now scores on a continuum of scores. If we collect data from a class on their first test, we might find the following ordered array of scores:

45, 48, 50, 50, 53, 53, 54, 56, 56, 57, 57, 59, 59, 60, 60, 61, 61, 61, 62, 63, 63, 65, 65, 65, 66, 66, 66, 67, 67, 68, 68, 69,

69, 70, 70, 71, 71, 71, 71, 72, 72, 72, 72, 72, 73, 73, 73, 74,
74, 74, 75, 75, 75, 76, 76, 77, 77, 77, 78, 78, 79, 79, 80, 80,
81, 81, 81, 82, 82, 83, 83, 84, 85, 85, 86, 87, 88, 88, 90, 93.

Even if we put this into a frequency distribution, it would still look
fairly formidable; that is, there are too many different numbers to deal
with (table 4-1). One way to avoid confusion is to lump a large number

TABLE 4-1 Frequency Distribution by Score on First Exam

score	f	score	f	score	f
45	1	66	3	79	2
48	1	67	2	80	2
50	2	68	2	81	3
53	2	69	2	82	2
54	1	70	2	83	2
56	2	71	4	84	1
57	2	72	5	85	2
59	2	73	3	86	1
60	2	74	3	87	1
61	3	75	3	88	2
62	1	76	2	90	1
63	2	77	3	93	1
65	3	78	2		80

of different scores together into groups of scores, thereby reducing the
number by putting them into intervals. We retain the information about
frequency but produce a more manageable set of scores to deal with.
Therefore, **grouping data** *is the process of clustering scores into a set
of intervals* (table 4-2).

TABLE 4-2 Grouped Frequency Distribution by Score on First Exam

Interval	f
45–49	2
50–54	5
55–59	6
60–64	8
65–69	12
70–74	17
75–79	12
80–84	10
85–89	6
90–94	2
Total	80

*How many intervals do we choose to put scores in,
and how wide should they be?*

It is easy to fall into the trap of making the number of intervals
as unwieldy as the original set of scores. Obviously the number of
intervals will depend upon their width. If I have scores that go from
20 to 100, and my intervals are 5 units wide, there will be 16 intervals;
however, if my intervals are 10 units wide, there will be 8 of them.
A good rule of thumb is never to make the intervals wider than the
maximum error you are willing to accept in placing values in that interval.
To use intervals of 10 dollars when referring to the price of cars, would
seem very picky and produce too many intervals to cope with. On the
other hand, using that same interval width to summarize prices of belts,
would effectively destroy all information since the majority would fall
in the 0–10 dollar range.

A second rule of thumb is to restrict the number of intervals to
a maximum of about 15. If there are more we run into problems of
internalizing the data at a glance. There are certainly times when fewer
intervals would suffice, but remember, the fewer the number of intervals,
the wider each has to be. A corollary to this is that if we include a
sufficient number of intervals, we retain the "feel" of the frequency
distribution. Having too few intervals loses much of the information
in the data (remember, in reporting data by grouping into intervals,
we ignore information about the individual scores).

A third rule of thumb is that the width of the intervals should
be in multiples of 5. It would ordinarily make little sense to use an
interval width of 7 or 16; this would only lead to confusion in the
table.

In short, the intervals should be few enough in number to allow
us to grasp the information within a few seconds of looking at the
table, yet large enough in number to retain information about the general
spread of scores.

*What happens to information about cumulative frequency
or cumulative percentage when data are grouped?*

The cumulative information is still useful, but you must
remember that the information is in terms of intervals, not individual
scores. Thus, to take our table of scores on the first exam, we might
report them as in table 4-3. Note that the cumulative frequency refers
to how many scores were attained up to and including that interval;
likewise, with percentage, we cumulate up to and including the given
interval, until we reach 100 percent.

TABLE 4-3 Distribution of Scores on First Exam

Interval	f	Cum f	Cum %
45–49	2	2	2.50
50–54	5	7	8.75
55–59	7	14	17.75
60–64	8	22	27.75
65–69	12	34	42.50
70–74	17	51	63.75
75–79	12	63	78.75
80–84	9	72	90.00
85–89	6	78	97.50
90–94	2	80	100.00
Total	80		

Why is there a "gap" between intervals in a grouped frequency distribution?

I certainly could have presented the intervals as 0-5, 5-10, etc., by merely following the convention that any score falling exactly on the value between two intervals is to be included in the higher interval. This, however, could lead to much confusion since not all people follow this convention. To avoid the problem, we make the intervals 0-4, 5-9, 10-14, etc. For instance, in the example of the scores on the exam, we assume that no one could receive 74.65 points. Therefore, if the score is 74, it falls in the interval 70-74. If it is 75, it falls in the interval 75-79. Hence any one of 5 values (e.g., 70, 71, 72, 73, 74) are included in an interval, and it retains its width of 5.

If, however, the data are continuous, we are implying that the true limits are, for example, 69.5 to 74.5. This is the rounding procedure we learned in elementary school. If the score falls in the "gap" between the intervals and the decimal fraction is above .5, we round up to the next interval. If it is below .5, we round down to the lower interval. If the decimal fraction is exactly .5, we round to the nearest even whole number. Thus, we are maintaining the true width of our interval as 5 even though it appears to be in units of 4.

Why bother to use "true limits"?

We distinguish true or real limits as opposed to apparent or stated limits to draw attention to the fact that while our table shows the limits of the intervals in whole numbers, for a continuous variable, we are actually stating the width of the interval to be .5 above and below. In general, *the **true limits** are values plus and minus one-half* *unit beyond the stated or* **apparent limits**. You should see that this statement extends the argument above such that no matter how many decimal places there are in the apparent limits, the true limits will be plus or minus .5 units beyond that place. For example, the true limits for the interval 5.0–9.9 are 4.95–9.95; the true limits for 5.00–9.99 are 4.995–9.995, and so on. True limits are important because we must use them, rather than stated limits, in computing the statistics we will be learning about in subsequent chapters. If not, we introduce an error into computations.

Can grouped data be graphed?

Clearly we can use circle graphs to represent percentages, but when the measurement is on an interval scale we can also plot the frequency by employing a particular form of the bar graph called a histogram (derived from histo = differentiation of cells, and gram = a specified number). *A* **histogram** *is a bar graph with the bars joined in order to indicate the interval continuum.* The bars represent frequency as in the qualitative case, but the base indicates the values on the interval scale. When data are grouped, it means we merely combine those values on the base when constructing the graph.

Which part of the graph represents the frequency in a histogram?

The horizontal line represents the interval scale and is called the **abscissa**, from the Latin root meaning to "cut or mark off" (a good portion of the foundations of mathematics were written in Latin). Hence, we mark off the horizontal line according to our interval scale. *The vertical line represents the frequency and is called the* **ordinate**, from the Latin word meaning "to order." Therefore, we are ordering the frequency according to the interval continuum. To avoid confusion between these two words, remember the mouth stretches horizontally when we say abscissa; it stretches vertically when we say ordinate.

How do we show true limits on a graph?

Generally we mark off the abscissa using the apparent limits. However, it is also common practice to arrange the bars in such a way as to bring them down on the abscissa at the point of the true limits, as in figure 4-1. Note that in quantitative graphing of histograms, we don't have to put the lines separating the bars in the graph because the abscissa is a continuum.

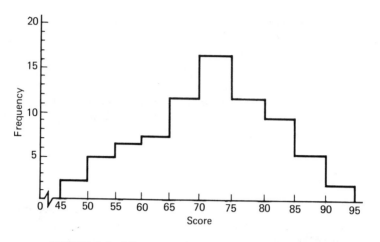

FIGURE 4-1 Histogram of students' scores on midterm

*Relative to each other, how long should the ordinate
and abscissa of a graph be?*

How a graph will look is determined by how the frequency
on the ordinate is marked and where you start your interval scale. Often
this decision may give a misleading impression. Notice that in graphs,
your eye tends to compare the relative difference of the heights, taken
from the top of the bar rather than from the abscissa, distorting the
figures, as in the two histograms illustrating the same data in figure
4-2.

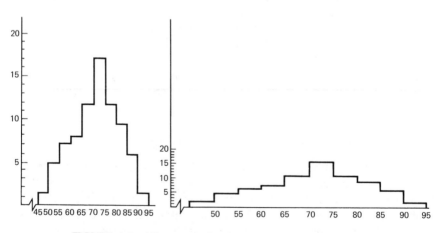

FIGURE 4-2 Histogram of scores distorting the ordinate
and abscissa

In general, try to keep the height of the highest frequency slightly less than the length of the horizontal axis. In addition, always try to start the frequency count, as well as the interval scale, at zero rather than arbitrarily beginning somewhere above zero. If cases do not begin to appear until a score somewhat to the right of zero on the abscissa, start the horizontal axis with zero and show the break in the continuum with a jagged line as in the graphs in figure 4-2.

What happens if the intervals are not equal?

At times, if we have a set of scores in which several are extreme and the others bunch up, or in which some parts of the distribution are sparse, we might want to make some of the intervals larger than others. Therefore, when reading histograms, it would be wise to think of the bars in terms of area rather than just height. For instance, if one interval is twice as wide as the others, it would be misleading if we indicated its frequency in the same way. Therefore, we must adjust the height of the bars to make the areas proportional to frequency. Take the example in figure 4-3. Notice that the interval 23–26 is twice as wide (4 units) as the interval 17–18 (2 units). Its frequency of 12

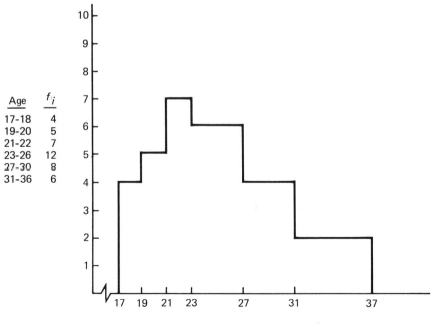

Age	f_i
17-18	4
19-20	5
21-22	7
23-26	12
27-30	8
31-36	6

FIGURE 4-3 Histogram for unequally grouped ages

is, therefore, spread over two intervals of width 2; hence the height of this bar is 6, not 12. In using this strategy, we have maintained the relative proportions of the frequencies in the areas of the bars rather than in the height of the bars.

How do you make a histogram if the end interval is "open-ended"?

If a set of categories for grouped data contains one or more intervals for which there is no boundary, we are in a dilemma when graphing these intervals. Although there is no clear tradition, one variation seems to be that in figure 4-4. You will notice that the open-ended

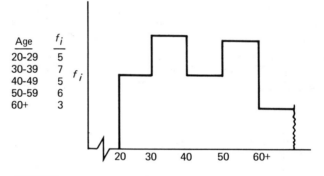

Age	f_i
20-29	5
30-39	7
40-49	5
50-59	6
60+	3

FIGURE 4-4   Histogram for open-ended categories by age

interval 60+ is pictured as a bar which is terminated with a jagged line. This line signifies much the same thing as when it appears on the abscissa; that is, it indicates that the bar continues to a further value, which in this case remains unspecified. In practice, however, this kind of graph is rarely seen and typically is left in table form.

Does a frequency polygon graph the same information as a histogram?

Yes, but rather than using the height of the bars to represent the frequency, we use a line that connects the heights of the frequency. Such a graph is called a frequency polygon. Think of it as a connect-a-dot cartoon you see in the comics, where the picture might end up looking like a mountain. *A* **frequency polygon** *is a histogram which has straight* *lines connecting the midpoints of the intervals at the height of their frequency.* The name comes from the fact that we are plotting frequency and we are using a polygon (i.e., a many-sided figure). Note in figure 4-5 that the frequency scale begins at zero, and the first point begins on the abscissa, half an interval below the starting point. Likewise the polygon reaches the baseline half an interval above the last interval. We do this to gain closure on our figure.

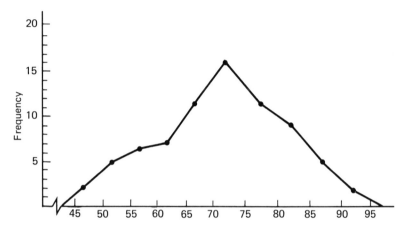

FIGURE 4-5   Frequency polygon of students' scores

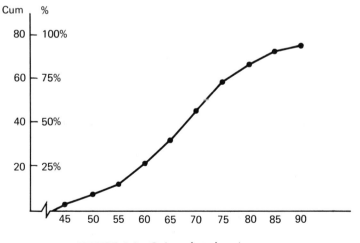

FIGURE 4-6   Ogive of students' scores

Can cumulative frequency also be shown graphically?

Yes. Quite often we want to graph a cumulative-frequency distribution. We can do this handily with a form of a graph called an ogive. *An* **ogive** *is defined as a cumulative-frequency polygon.* The word ogive comes from the French word meaning an S-shaped curve. In general, any time we have a distribution with the majority of the scores in the center, the cumulative frequency polygon or ogive will have the form of an elongated S. The horizontal axis is still the interval scale, and the vertical axis can be marked off in cumulative frequency or cumulative percentage. Note in figure 4-6 the first point starts at zero frequency and ends at N or 100 percent. This is generally true for an ogive.

Are there other kinds of graphs?

Yes. There is the graphing of a single variable over time. You are probably most familiar with this in the form of the stock listings

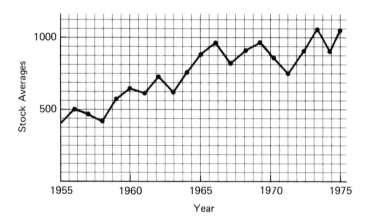

FIGURE 4-7   Graph of hypothetical averages 1955–1975

in the financial pages or a doctor's chart at the end of a patient's hospital bed. Most often we put the value of the variable on the ordinate (vertical axis) and draw a line connecting values across time which are marked off on the abscissa (horizontal axis). This is sometimes referred to as **a trend line** and is heavily used in economics to graph trends over time (fig. 4-7).

In a slightly different context, there is a type of graph which connects scores with lines. It appears predominantly in experimental work and illustrates the differences between values in one or more groups (fig. 4-8). This type of graph is useful in presenting a quick visual interpretation

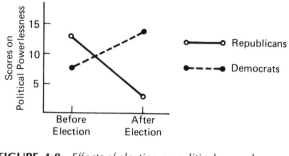

FIGURE 4-8   Effects of election on political powerlessness
by party

of differences. Note here it appears Republicans became less powerless
while Democrats became more powerless after the elections.

> *Can quantitative data be summarized without using
> tables or graphs?*

Yes, Since all data form some sort of distribution, we are
able to describe the relevant characteristics of these distributions.
In the next chapters we study two kinds of tools. First, we have measures
of central tendency to indicate how the scores congregate around the
center of a distribution. Second, we have measures of dispersion to
indicate how the scores spread around those central tendencies.

How Much Do You Remember?

New Words

Grouped data
True limits
Apparent limits
Histogram
Abscissa
Ordinate
Frequency polygon
Ogive
Trend line

New Symbols

cum *f*

Did You Ever Wonder?

1. What information is lost by grouping individual values when talking about
national average income for families?

2. Why do bar graphs have disconnected frequencies while histograms don't?
3. Which causes the greater amount of misinterpretation in graphing: stretching the abscissa or stretching the ordinate?

Want to Know More?

1. ANDERSON, T. R. and M. ZELDITCH, JR. *A Basic Course in Statistics.* 3rd ed. New York: Holt, Rinehart, and Winston, Inc., 1975. Chapter 4.
2. HUFF, DARRELL. *How to Lie with Statistics.* New York: W. W. Norton & Company, Inc., 1954.
3. MORONEY, M. J. *Facts from Figures.* Baltimore: Penguin Books, Inc., 1951.
4. SCHMIDT, M. J. *Understanding and Using Statistics.* Lexington, Mass.: D. C. Heath & Company, 1975. Chapter 2.

Central tendency

BALANCING
THE TEETERBOARD

5

What is the meaning of the term central tendency?

In looking at a graph of the frequency distribution of a given variable, two things are usually noticeable: first, the values of the variable tend to group around a central value or values; and secondly, they also disperse around that value in a specific way. *Describing the central point(s) around which values in a distribution cluster is what we mean by measures of* **central tendencies**.

Is central tendency the same as an average?

By definition, yes. Think of it this way. When you were in high school taking a physical education class, you probably had to do some running with the rest of the gym class. Did you ever notice that some people were always faster runners than others, and some were invariably slower? The bulk of the people, however, were always in the middle. If we were in the business of understanding this phenomenon, we would want to have at our command ways of describing the speed of most of the runners clustered somewhere between the fastest and the slowest. The use of averages as measure of central tendency would allow us to do this.

43

*You use the word "averages." Does this mean that there is
more than one average?*

Yes. There are three common but distinct meanings of the
word average. No doubt, this may lead to some confusion. Indeed,
in both popular and technical literature people may use an inappropriate
measure of central tendency, that becomes a very potent weapon for
manipulating statistics. We hope to help you avoid being misled in this
way. As an opening statement, however, let's just say that the three
kinds of "averages" are: *the* **mode**, *which means the most frequent value
in the distribution; the* **median**, *which gives information about the value
of the middle position in the distribution; and the* **mean**, *which indicates
a point around which the values in the distribution "balance."*

Does each one of these measures have its own symbol?

Yes. The mode is most generally noted by Mo or just the
word mode. The median is abbreviated Md, and the mean is indicated
by a bar (‾) over the letter representing the variable. The letters most
often used are the capital X or Y. Thus, the mean of X is $\bar{X}$.

You will also remember that we made the distinction between using
English letters for samples and Greek letters for the population. In a
population, therefore, the mean is referred to as mu (μ), which is the
Greek symbol corresponding to *m*. The mode and median generally retain
the English letters even when referring to the population. Of course,
when you get right down to it, the symbols are arbitrary. Tradition
has perpetuated their use, and like any language, they have survived
over many generations.

Is the mode a frequency or an interval?

If we look at the frequencies associated with each category
or interval on a given scale, then the mode is merely that position which
has the greatest proportion of the total frequency associated with it.
In other words, it is the value or category whose position most frequently
occurs.

*Do we use the mode when talking about nominal
and ordinal scales?*

We do. The mode is best suited for talking about nominal
kinds of scales. We use it to indicate the nominal category that is the

most prevalent. There are more freshmen in college than any other year; there are more X-rated movies than any other kind; more persons in United States who identify themselves as Catholics than identify with any other single religious group.

However, the concept of mode is equally applicable to ordinal-level data. In the army, men often take a census to see if there are more enlisted men, noncommissioned officers, or commissioned officers on the base during the weekend. Rank is ordinal-level data, and knowing this information is especially important, as the ability to leave the base is dependent upon how many commissioned officers are present. The drawback in using the mode with ordinal-level data, however, is that it excludes use of extra information about the central tendency, namely, the order in which the categories occur.

Is the median appropriate for ordinal data?

If I rank students in the class and then give the median rank, that median rank is a function of the number of people N and not of the values or scores or characteristics on the basis of which ranks were determined. The median for ordinal data thus conveys no real information unless there are tied ranks. Even then, it conveys very little information. The notion that the median is especially useful with ordinal data is a common fallacy. What we really want to know when using the median is the value associated with the position in which half the scores are more extreme and half are less extreme. This, of course, is impossible with ordinal data since there are no values, only order of positions. Therefore the median is used with interval data.

What does it mean to speak of a median for grouped interval data?

The median is the *value* of the middle position. To find this value for grouped data, we determine the category within which the middle position is contained. Next, we add to the lower limit of that category (i.e., lower true limit) that proportion of the way into the category that the middle position falls; to make sure we are adding values and not just positions, we multiply that proportion by the width of the category (i.e., highest minus the lowest possible value in the category). This gives us the value associated with the middle position; and hence the formula

$$\text{Median} = X_{ll} + \frac{N/2 - \text{cum } f_{ll}}{f_i} \, i.$$

The first term is the value of X at the lower limit of the category containing the median. The second term is that proportion of the category that falls below the median ($N/2$ is the middle position minus cum f_{ll}, all the positions up to the category, divided by f_i, the number of positions in that category). The final symbol i is the width of the category.

Of course, if the data are not grouped, the median for interval level is just the value of the middle score; that is, the value at the $(N + 1) \div 2$ position. If the number of values is odd, for example, the seven scores 3, 4, 6, 7, 9, 9, 11, the median would be the value of the fourth score, or 7. If the number of positions is even as in the six scores 3, 5, 5, 7, 9, 11, the median would be the value at the $(6 + 1) \div 2$, or 3.5, position. This would be the average of the values at the third and fourth positions, or $^{(5+7)}/2 = 6$.

Is the median for grouped data always in the middle group?

No. It doesn't have to be the middle group. It only has to be the group that contains the middle score. Think of it this way. If you went shopping and accidentally parked your car at what seemed an inconvenient five blocks away from the store, you might be curious to compute the average distance that all the people who had parked in those five blocks had to walk to reach the store. If, for example, they were equal-sized blocks with street parking near the store, each

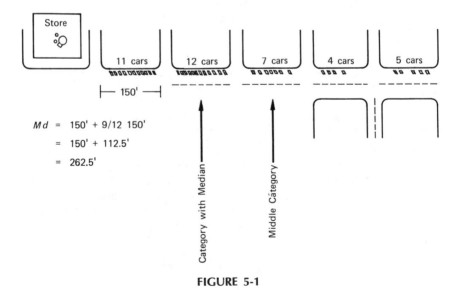

Md = 150' + 9/12 150'

     = 150' + 112.5'

     = 262.5'

FIGURE 5-1

having a different number of cars occupying the spaces, you might use the distance from the car in the median position as a measure of the distance the "average" person had to walk in order to reach the store. It should be obvious that the median position may not fall in the third block, which would be the middle section.

Suppose the number of cars in each block were 11, 12, 7, 4, 5, respectively. The total number of cars would be 39 and the middle position 20. Obviously, the 20th car is not in the third block, but in the second block, which contains 12 cars. The distance would be difficult, if not impossible, to measure by yardstick from the store. However, we may know the length of one block. To this, we merely add $9/12$ the distance of another block. The reason for this is that the 20th position is in the second block and it is $9/12$ of the way into this second block. Thus, $1 \frac{9}{12}$ blocks is a measure of the median car's distance from the store (fig. 5-1).

Is the median the only adequate central tendency measure for interval data?

Adequate is not a very precise word. The median supplies information about the value associated with a particular position relative to all other positions. In particular, it gives information about the value of the middle position. Unfortunately, we often want to take into account the values associated with all the positions, not just the middle position, when computing an average. In order to do this, we may utilize some notion of the balance of these values, which is why we more often use the mean as an average in interval-level distributions.

Why is the mean a "balance" point in a distribution?

The mean has two very important properties. The first I have chosen to call the teeter-totter principle. The teeter-totter principle says that the mean is a number such that the sum of the differences of all values from that mean is equal to zero. This is why the concept of balance enters in. Think of it this way. Have you ever been in a park on a seesaw or teeterboard with a friend whose weight is quite different from yours? If he was much heavier, no doubt you were left hanging in the air as he pushed down. If you were heavier, the opposite was true. Observe that either the balance point must be moved in order for you to play on the teeterboard or more weight must be added to the lighter side to equal the weight opposite it. In short, in order to balance on a fulcrum, the values of the weights times the distances

Think of the mean as a fulcrum. In order to balance, the
values of the weights times the distances from the balance
point must be the same on both sides

from the fulcrum must be the same on each side. In a distribution
of values on an equal-interval scale, the sums of the distances from
the mean must be the same on each side of that mean. That is,
$\Sigma(X_i - \bar{X}) = 0$.

Why does the mean as a central value represent a "minimum"?

The second property of the mean states that the sum of the
squared deviations from the mean will be less than the sum of squared
deviations from any other point. If we subtract the mean from each
value in the distribution, squaring these differences and summing them,
this value (called the sum of the squares) will be less than from any
point other than the mean. Hence, $\Sigma(X_i - \bar{X})^2$ yields the minimum
sum of squared deviations. The fact that this sum is less than when
taken from any other point implies that the mean is the best guess
of the values in the distribution using this "least squares" criteria; that
is, the overall error (squared deviation) from the mean as a guess for
all scores will be less than when using any other value in the distribution.
This will take on added significance in subsequent chapters.

How is the mean computed?

The computation of the arithmetic mean is ridiculously simple. Just
add up all the possible values and divide by the total number of values.

Hence, the formula

$$\bar{X} = \frac{(X_1 + X_2 + X_3 + \ldots + X_n)}{N} = \frac{\Sigma X_i}{N}$$

(obviously, if we are speaking of a population it would be $\mu = \Sigma X_i / N$).

Does the same computational process for the mean hold true when the data are grouped?

You will remember that when we group data, a certain amount of information about each value is lost in favor of providing a tidier overall representation of the distribution. When computing the mean, we have to assume that each value in the group is represented by the midpoint of the category. We then merely multiply each midpoint value by the frequency in the interval. The last step is to divide, as before, by the total N. Hence the formula:

$$\bar{X} = \frac{\Sigma f_i \cdot midpoint}{N} = \frac{\Sigma f_i m_i}{N}$$

You mention the arithmetic mean. Is there more than one kind of mean?

Yes. In particular there are four interpretations of the mean. They are the arithmetic, geometric, harmonic, and quadratic means. Computation of the arithmetic mean is mentioned above. The geometric mean is defined as the Nth root of the product of all the values $\sqrt[N]{X_1 X_2 X_3 \ldots X_n}$. The harmonic mean is the total N divided by the sum of the reciprocals of the values $N/\Sigma(1/X_i)$. The quadratic mean is the square root of the quantity: the sum of the values squared divided by N or $\sqrt{\dfrac{\Sigma X_i^2}{N}}$. The arithmetic and weighted arithmetic mean are used most extensively in social science, and the latter three in physical science. Hence, the most important interpretations of the mean in elementary social statistics are the arithmetic and the weighted arithmetic methods.

What is a "weighted" arithmetic mean?

This term describes the situation where we may have more than one group, each with its own respective frequency f_i. Hence, it is necessary to weight each group's mean by the number of objects in that group to get an overall mean. The mean for grouped data can be interpreted as a specific case of a **weighted mean**. We have to multiply the value of the midpoint of each category by however many of those values that exist. Hence the formula $\Sigma(f_i \bar{X}_i)/N$. This gives us the proper weighting in computing the average. You may see some authors using n_i in place of f_i. The meaning is the same, however, and they can be used interchangeably.

Is the mean really the best "average"?

No, not at all. Each measure of central tendency has its own specific purpose. The *mode* is the best spot-check for frequency. The *median* best describes the value of a position, and the *mean* utilizes information about the values associated with all positions. With interval data, the mean gives us the advantage of using all the information available.

How do you know which measure of central tendency to use?

Returning to our discussion of the frequency polygon, you will remember that more often than not the distribution is not perfectly symmetric. The largest chunk will appear somewhat to the left or right. When this occurs, we say that the distribution is **skewed**. *If the distribution is negatively skewed, the bulk of the values fall toward the right side; if it is positively skewed, the largest proportion of the values fall to the left side.* Skewness affects different measures of central tendency in different ways. By adding an extreme score, we move the median somewhat in the direction of the added score. The mean, however, is more sensitive to the values of the distribution; hence, the mean is also moved in the direction of the added score, but the change is greater the more extreme the value of that added score.

Why is the mean more sensitive to extreme scores than other measures of central tendency?

Since the mode does not take into account values, the effect of an additional extreme score is minimal. The median is the value of one position; therefore, adding an additional position makes little difference there, too. However, because the mean depends on all values in the distribution, an additional extreme value will have a more significant effect. For example, in a positively skewed distribution the mode has the lowest value, the median is next lowest, and the mean has the

The mean is extremely sensitive to the values of the
distribution; hence, the mean is moved in the direction
of skewness, but in a much more drastic manner

highest value. The difference between the mean and median is a crude
measure of the skewness of a distribution.

If you were the mayor of a city applying for federal poverty program
grants and you knew that income is positively skewed, you would most
certainly report the mode as the average income. If, on the other hand,
you were a member of the chamber of commerce, you would probably
want to report the mean as the average income.

*What you are trying to say is that people use the term average
without specifying which average, in order to justify
their own ends. Right?*

It's probably not all that conscious an effort on most people's
part. More than likely, people don't know exactly what measure of
central tendency is best suited for what they are trying to describe.
However, there are a few unscrupulous characters who would purposely
mislead. Darrell Huff, in his book *How to Lie with Statistics*, gives
us an excellent example of the manipulation of statistics in describing
how corporation owners might report the mean rather than modal income,
in order to indicate a higher "average" salary. The point here is never
accept the average as indicating one thing when, in fact, it may mean
something quite different. A general caution in summarizing distributions:
if the interval-level variable is extremely skewed, use the median as
a measure of central tendency; otherwise, use the mean.

*In a bell-shaped curve, isn't the mean the same as
the median and mode? If so, what advantage
does using the mean have?*

You're right. In a symmetrical distribution, the value of the
mean, the median, and the mode are all the same. However, the mean
has two features that make it generally useful whatever the shape of

the distribution. First, it is the only measure of central tendency that takes all values of the distribution into account. Secondly, it provides the mathematical link to a powerful measure of how values are distributed about the mean of a distribution (whether the distribution is symmetrical or not). Thus, it serves as a logical step in leading us from an understanding of central tendencies to a knowledge of variations. The study of variation is the subject of our next chapter.

How Much Do You Remember?

New Words	*New Symbols*
Central tendency	Mo
Mode	Md
Median	$\bar{X}$
Mean	μ
Weighted mean	
Skewness	

Did You Ever Wonder?

1. Why does the U.S. Government report the "median" national income while reporting the "mean" educational level?
2. Why is the mean more "sensitive" than the median to changes of values in a frequency distribution?
3. What is meant by the term "average American?"

Want to Know More?

1. PEARSON, KARL. *The Grammar of Science.* 3rd ed. New York: Meridian Books, 1957.
2. WALLIS, W. A., and H. V. ROBERTS. *The Nature of Statistics.* New York: Collier Books, 1962.
3. REICHMAN, W. J. *Use and Abuse of Statistics.* New York: Oxford University Press, Inc., 1962.
4. WEINBERG, G. H., and J. A. SCHUMAKER. *Statistics: an Intuitive Approach.* Belmont, Calif: Wadsworth Publishing Company, 1962. Chapter 2.

Variation

DEALING WITH DEVIATIONS

6

What does variation mean?

We defined a variable as meaning that trait which can take on a range of values. When we talk about variation in an interval-level distribution, we speak of the arrangement or spread of values that the variable takes in the distribution.

Do all distributions, including those of qualitative attributes, show variation?

All show variation. However, only quantitative distributions utilize the values of the variables to compute a measure of variation. Qualitative data or attributes show variation in classification and in the frequency or proportion (relative frequency) of the various classes. But for the purposes of this chapter we will discuss measures of dispersion only for quantitative data.

Is there any quick check of the spread of a distribution that doesn't involve a lot of computation?

Yes, the **range**, *defined as the highest score minus the lowest score*, is specifically suited for this purpose. When dealing with grouped data, an estimate of the range can be computed by subtracting the midpoint of the lowest interval from the midpoint of the highest interval. The range is often used along with the mode to provide a quick reference to the central value and spread of a distribution.

*Does the range really tell very much about
the shape of the distribution?*

The range does not provide information about the spread of scores around the central tendencies in that distribution. It is possible that a distribution with one hump in the frequency polygon will have the same range as a distribution with several humps (fig. 6-1). Intermediate

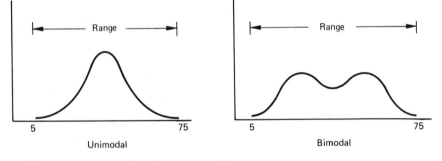

Unimodal   Bimodal

FIGURE 6-1

ranges, such as the **interquartile range** (Q), do not indicate the exact shape of the distribution but they are less dependent than the entire range or the peculiarities of the most extreme cases. To find the interquartile range, which is the range including the middle 50 percent of the cases, we first locate the 75th percentile and the 25th percentile (in a manner parallel to that by which we locate the 50th percentile with our formula for the median [see last chapter]); we then compute the difference between the 75th and 25th percentile to find the interquartile range. If the data were grouped, we would merely follow the formula for finding the percentile with grouped data.

*Is there a measure of variation that makes use of all
the values of the distribution?*

We might subtract the mean from each value in the distribution and find the average (mean) of these deviations. This result would take into consideration the first property of the mean (i.e. the teeter-tottor principle). In the last chapter, however, we learned that this principle, by definition, indicates a value of zero. One way to modify this resulting zero and still make use of the property is to take the absolute value of these deviations (remember that the absolute value merely means omitting the sign in front of the number so that all values are positive). Therefore, the sum of the absolute values of those deviations will be $\Sigma|X_i - \bar{X}|$. This tells us something about the spread of the distribution. However, we prefer to speak of deviations per observation and therefore divide this sum by the total number of cases (N). Hence, $\dfrac{\Sigma|X_i - \bar{X}|}{N}$ is called the **mean deviation** (M.D.). A larger value indicates a greater spread in the values of the distribution.

Is the variance a "better" measure of dispersion than the mean deviation?

Yes. We will see in subsequent chapters that the concept of **variance** is important in describing and testing relationships between two or more variables. However, the variance also provides an important conceptual link in descriptive statistics. Rather than taking the absolute value of deviations, the variance is computed by squaring each difference. Because a positive or negative number times itself always results in a positive product, we eliminate the problem of negative values. More important, we utilize the second property of the mean; that the sum of the squared deviations around the mean is a minimum value. Thus, the sum of the squared deviations divided by N gives us an appropriate measure of squared deviations per case. $\Sigma(X_i - \bar{X})^2/N$ is the formula defining the variance. Often we see $(X_i - \bar{X})^2$ replaced by x_i^2. Thus the variance may be represented by $\Sigma x_i^2/N$.

How is the standard deviation to be interpreted?

Think of the mean as a stable reference point in a distribution. Each of the other values represents deviations that act as forces causing that case to be different from others in the group. Since each observation is independent of the others, each deviation represents a unique force. To gain some understanding of these forces, we must somehow "combine" them. It was suggested above that one mathematically rigorous method is to square the deviations in computing the variance. However, squaring the deviations also increases their magnitude, and therefore the sum of the squared deviations become correspondingly larger. The standard deviation attempts to minimize this increase by taking the square root of these squared deviations. We thus return to our original units. Algebraically, the standard deviation is merely the square of the variance and defined by the formula

$$s = \sqrt{\frac{\Sigma(X_i - \bar{X})^2}{N}}.$$

Actually, the concept of standard deviation was developed before the concept of variance and was denoted by the Greek symbol σ. Thus, when the concept of variance appeared, it was denoted by σ^2. We still use these symbols today when describing a population variance and standard deviation. But in a sample we denote them by their English letter equivalents and simply use s and s^2 respectively. In practice, however, authors often interchange s and σ when describing the standard deviation. Frequently, we see $s = \sqrt{\frac{\Sigma(X_i - \bar{X})^2}{N}}$ rather than $\sigma = \sqrt{\frac{\Sigma(X_i - \mu)^2}{N}}$ used as the common formula to indicate the standard deviation in descriptive statistics. In Chapter 12, however, we will see an important distinction between using s as a descriptive statistic and using it as an unbiased estimate of the population variance.

Is there an easier way to compute the variance and standard deviation without having to subtract the mean from each value?

There is, but to recognize that the computational formula is essentially the same, you must remember one little fact from high school math. This will be one of only two derivations you find in the book, but it is here for a reason, so pay close attention. First, look at the two formulae for the variance.

$$\frac{\Sigma(X_i - \bar{X})^2}{N} = \frac{\Sigma X_i^2 - \frac{(\Sigma X_i)^2}{N}}{N}$$

Definitional   Computational

If we take the numerator of the first, or definitional, formula, we should recognize it as a special form of the problem of expanding the difference of two terms: $(a - b)^2$. The summation Σ merely means that we put that sign in front of each of the terms of the expansion. Possibly you had a tough time with the rules for expansion, but try this simple acronym:

FOIL. FOIL means that you take the *F*irst term of the expansion and square it, and so we get a^2. We then take the *O*uter and *I*nner terms and add together their product; here we get $-2ab$. Finally we take the *L*ast term and square it to get b^2. Thus, the expansion becomes $a^2 - 2ab - b^2$. This means that the numerator for the first formula of the variance is $\Sigma(X_i^2 - 2X_i\bar{X} + \bar{X}^2)$. If we merely combine and reduce this term, we can easily see that it becomes the numerator for the second or computational formula for the variance:

$$\Sigma(X_i^2 - 2X_i\bar{X} + \bar{X}^2)$$
$$= \Sigma X_i^2 - \Sigma 2X_i\bar{X} + \Sigma \bar{X}^2$$

By taking the constant 2 out of the summation in the middle term and remembering that in the last term the sum of any constant is N times that constant, we get

$$\Sigma X_i^2 - 2\Sigma X_i\bar{X} + N\bar{X}^2$$

Also, remembering that ΣX_i is $N\bar{X}$ (i.e., $\bar{X} = \Sigma X_i/N$), we get

$$\Sigma X_i^2 - 2N\bar{X}\bar{X} + N\bar{X}^2$$
$$= \Sigma X_i^2 - 2N\bar{X}^2 + N\bar{X}^2$$

Combining terms we get

$$\Sigma X_i^2 - N\bar{X}^2 = \Sigma X_i^2 - N\left(\frac{\Sigma X_i}{N}\right)^2$$

Now dividing this whole term by N as in the original formula, we have the second, or computational, version

$$\frac{\Sigma X_i^2 - N\dfrac{(\Sigma X_i)^2}{N^2}}{N} = \frac{\Sigma X_i^2 - \dfrac{(\Sigma X_i)^2}{N}}{N}$$

In computing the standard deviation, we take the square root of these formulae:

$$s = \sqrt{\frac{\Sigma(X_i - \bar{X})^2}{N}} \quad \text{or} \quad s = \sqrt{\frac{\Sigma X_i^2 - \dfrac{(\Sigma X_i)^2}{N}}{N}}$$

The reason for leading you through this derivation is not to present you with the mathematics but to show you that this, like any other computational formula, is merely a result of the expansion and combination of terms in the original formula—nothing more. It should be remembered therefore that, whenever a strange computation appears, ultimately it refers to the original statement but merely in another form. Keep this in mind as you move on in your study of statistics.

The utility of this computational formula is that we can compute the standard deviation without having to subtract the mean from each variable. Indeed, if we square each value and sum them, sum the values and square that sum dividing it by N, subtract the second value from the first, and finally divide the N, we accomplish the same task.

What is the procedure for finding the variance
if the data are grouped?

If the data are grouped, we are dealing with frequencies in each category. Each value is assumed to be at the midpoint of the category in which it falls. Therefore, we can use both the definitional and the computational formulae for the standard deviation and variance, replacing the value of X with the midpoint of the interval times the frequency in the interval:

definitional formula:

$$\frac{\Sigma(X_i - \bar{X})^2}{N} = \frac{\Sigma f_i(m_i - \bar{X})^2}{N} \qquad \text{when data are grouped}$$

computational formula:

$$\frac{\Sigma X_i^2 - \dfrac{(\Sigma X_i)^2}{N}}{N} = \frac{\Sigma f_i m_i^2 - \dfrac{\Sigma f_i m_i^2)^2}{N}}{N} \qquad \text{when data are grouped}$$

The standard deviation is merely the square root of these formulae.

How does the standard deviation help us describe
individual scores?

Remember that the standard deviation is a measure of the "average" spread of scores around the mean. But what if we want to isolate an individual score from a distribution and compare it with some other score in the same or another distribution? Means and standard deviations won't help in comparing individual scores; however, we can use them in conjunction with the individual score to create a very meaningful comparison. For example, we could "standardize" the individual score by describing its distance from the mean in terms of standard deviation units. That is, we would measure how many standard deviations away from the mean the individual's score falls. This is expressed by subtracting the mean from the score $(X - \bar{X})$ and determining how many standard deviations this distance comprises. Thus, we divide $(X - \bar{X})$ by s, to compute the **standardized score**.

How does standardizing help compare scores?

If we say John scored 80 on a test in one class and Jeannie scored 70 in another class, we would have no reference point for comparison. On the other hand, by comparing scores from two distributions in terms of the number of standard deviations from the mean, we can compare the *relative* position of the scores in the same or different distributions. The score is expressed relative to the mean and measured in standard deviation units. Think of this analogy. You know people in two cities, each of whom you would like to visit; however, you have only one day to do it. In order to decide on your destination, you might compare the distance of each trip. This, however, would do little good as one trip might involve city traffic while the other might be on a freeway. Therefore, you use a unit by which to standardize—the average speed you can travel. Your comparison becomes meaningful now, in that you are dividing distance by speed to determine the number of hours of travel time required to reach each destination. This consideration is more important anyway, since the problem you must solve is the time constraint.

In the same way, we compare the number of standard deviation units from the mean rather than the "raw" distance from the mean in comparing individual scores. The point is that now we can compare scores from distributions with different means and standard deviations by referring to the relative position rather than the raw score. *We call this standardized score* a *z* **score**. $z = (X_i - \bar{X})/s$. This then, is another new symbol in the vocabulary of statistics, defined in terms of the raw score, the mean, and standard deviation (fig. 6-2).

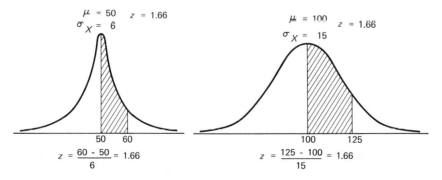

$\mu = 50$
$\sigma_X = 6$
$z = 1.66$

$\mu = 100$
$\sigma_X = 15$
$z = 1.66$

$$z = \frac{60 - 50}{6} = 1.66$$

$$z = \frac{125 - 100}{15} = 1.66$$

FIGURE 6-2

Why is the z score a measure of relative position in distributions?

This is one of the primary reasons why the special theoretical curve called the normal distribution is so important. Its exact shape resembles a roller-coaster or symmetrical bell-shaped curve. In many empirical distributions, the graphic representation of the frequency distribution resembles this normal curve. For such a distribution, we may determine from the z score the proportion of cases falling between any two points in the distribution.

Specifically, this relationship between the z score and proportions of a normal curve allows us to make statements such as the following: Roughly 68 percent of the values in a given distribution will fall between +1 and −1 standard deviation away from the mean; roughly 95 percent of all frequencies will fall between +2 and −2 standard deviations; and roughly 99 percent of all frequencies will fall between +3 and −3 standard deviations (fig. 6-3).

We can assume this precisely because we define the shape of the normal curve, in part, by the standard deviation. We will look at this precise relationship more completely in the chapters on inferential statistics.

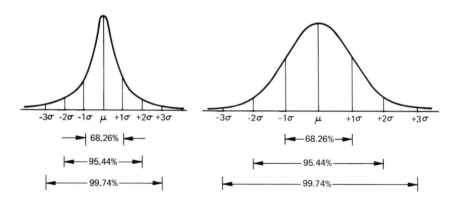

FIGURE 6-3

How was the relationship between frequency and the normal curve discovered?

The normal curve was originally called the curve of error, and stems from the early eighteenth century when skillful astronomers were finding that in observing stars, there was a distribution of errors in the measurement of the position of heavenly bodies. The farther away the outcome from the expected or predicted value, the less frequent the observed error.

This led some distinguished mathematicians to believe that the normal curve, or curve of error, was a fundamental law of the universe. Today, we know there are other kinds of distributions, but the normal curve is still very important for social research. Many traits such as height and weight as well as cognitive processes like I.Q. and strength of attitude are normally distributed. But most important, the normal curve is the basis for much of inferential statistics, as we shall soon discover.

Do z scores have a distribution?

Yes. If I have a set of normally distributed scores, subtracting their mean and dividing them by their standard deviation will not change the shape of their distribution, but it will translate the value of the mean and standard deviation, of their distribution. Thus, if we standardize all the scores (i.e., if we compute z scores) in a normal distribution, no matter what the values of that distribution, we create a translated or standardized distribution. *We call this z distribution a* **standardized normal distribution** *having a mean of zero and a standard deviation of one.* The mathematical symbol for this (a new one) is $\mathcal{N}(0,1)$. That is, the script $\mathcal{N}$ indicates a normal distribution; the first number in parentheses is the mean, and the second number is the standard deviation.

Why does the standard normal distribution have a mean 0 and the standard deviation 1?

There is nothing complicated about this. Look at the definition of the standard score: $z = (X_i - \bar{X})/s$. In order to find the mean of these scores, we sum them and divide them by how many there are

$$\bar{z} = \frac{\Sigma z_i}{N}$$

But we know by the first principle of the mean (teeter-totter principle) that the sum of the deviations around the mean equal zero, and therefore $\Sigma z_i = 0$. This is the first property of the standard normal distribution:

1. $$\bar{z}_x = \frac{\Sigma (X_i - \bar{X})/s}{N} = 0$$

In order to find the variance of z scores, we merely take the difference of each z score from its mean, square it, sum these up, and divide by the total N. Symbolically it is

$$\sigma_z^2 = \frac{\Sigma (z_i - \bar{z})^2}{N}$$

In looking at this, you should recall what we have just stated, namely, that the mean of z is zero. Thus, our formula for the variance of z is really just

$$\sigma_z^2 = \frac{\Sigma z^2}{N}$$

Now in order to reduce this, just take what you know about z and square it. Thus, we merely expand the numerator and cancel everything.

$$\frac{\Sigma z^2}{N} = \frac{\dfrac{\Sigma (X - \bar{X})^2}{s^2}}{N} = \frac{\dfrac{1}{s^2} \cdot \dfrac{\Sigma (X_i - \bar{X})^2}{1}}{N}$$

$$= \frac{\dfrac{1}{\dfrac{\Sigma (X_i - \bar{X})^2}{N}} \cdot \dfrac{\Sigma (X_i - \bar{X})^2}{1}}{N}$$

$$= \frac{\dfrac{N}{\Sigma (X_i - \bar{X})^2} \cdot \dfrac{\Sigma (X_i - \bar{X})^2}{1}}{N} = \frac{N}{N} = 1$$

If the variance of z is 1, then surely the standard deviation of z is also 1 since it is the square root of the variance. This is the second principle of the standard normal distribution.

2.
$$\sigma_z = \sqrt{\frac{\Sigma (z_i - \bar{z})^2}{N}} = 1$$

Again, you are not being taken through the mathematics to become competent in deriving formulae, but rather to understand that all we are doing is expanding terms and reducing them.

*How does knowledge of the standard normal distribution
help compare scores?*

First, we have said that standardizing a value allows us to
refer to its relative position in the distribution. In a normal distribution,
this relative position can be precisely expressed in terms of the proportion
of cases falling above or below that point. Because in a normal distribution,
the mean is equal to the median that is equal to the 50th percentile,
we can express these relative positions in percentiles (see Fig. 6-4).
A score of 115 is as far above the mean as 85 is below. That is, they
are both one standard deviation away:

$$z = \frac{115 - 100}{15} = 1 \qquad z = \frac{85 - 100}{15} = -1$$

Since 68.23% of the values fall between ±1 standard deviation, we
compute our percentiles from this figure. In the case of the negative
z value, we subtract that percentage from 50 percent to find the percentile;
in the case of a positive z value, we add that percentage to 50 percent.
Therefore, the two percentiles in our example are 15.87 (50 minus half
of 68.26 percent) and 84.13 (50 plus half of 68.26 percent), respectively.

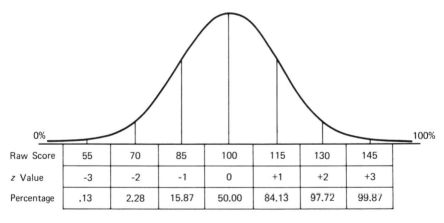

Raw Score	55	70	85	100	115	130	145
z Value	-3	-2	-1	0	+1	+2	+3
Percentage	.13	2.28	15.87	50.00	84.13	97.72	99.87

FIGURE 6-4

*Can a cumulative percentage be computed if the z score
is measured in fractions of a standard deviation?*

If you computed a z score of 1.5, for example, you might
be tempted to interpolate by taking half the percentage between 1.0

and 2.0 standard deviations and add it to the percentage for 1.0. This, however, would lead to difficulty. Because the relationship between the z score and the proportion of cases between it and the mean is not linear, we cannot find values this way. Indeed, by merely looking at the difference in percentage in figure 6-4, between ± 1.0 and ± 2.0 standard deviations, we can tell that the percentage of cases in the interval between $z = \pm 2.0$ is not twice that of $z = \pm 1.0$.

Fortunately, a table showing the proportion of the area of normal curve (i.e., the proportion of cases) that falls between the mean and selected z scores has been available since a statistician named Sheppard derived it in 1903. Since then, one has had merely to look up the percentage corresponding to a particular z value to determine the proportion or percentage value associated with it.

How is a percentile derived from a table of z values?

The table (see appendix IV) is arranged so that the z value is listed down the left-hand column to the nearest tenth. Across the top is the hundredths place. For example, a z value of 1.66 is found in the row marked 1.6 and the column marked 6 across the top. The value found at the intersection of that column and row is the decimal equivalent of the percentage of frequency between the mean and the point. Thus, we know that a z value of 1.66 indicates .4515, or 45.15 percent, of the cases in the distribution are between the mean and that point. The cumulative percentage below that value of 1.66 is, therefore, 50 percent + 45.15 percent = 95.15 percent (fig. 6-5). For a negative z value, we would *subtract* the percentage shown in the table from 50 percent.

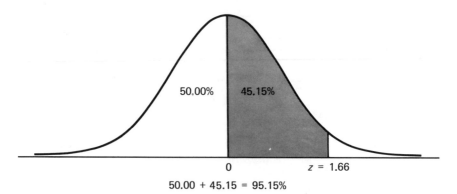

50.00%   45.15%

0   z = 1.66

50.00 + 45.15 = 95.15%

FIGURE 6-5

How can we use a z table to find the percentage between two z scores?

Consider the normal curve as a roller-coaster. Now suppose a contractor were interested in how much of the total supply of wood must be used to build certain sections of frame under the track. If the size of each section were one standard deviation unit, he might want to know what percentage of the total stock of wood he would use in the area of, for example, .38 to .78 sections out from the middle of the track. Assuming the track to be somewhat normal in shape, he would merely find the percentage attached to each z score and subtract them in order to solve the problem.

Suppose a contractor were interested in how much of the total supply of wood must be used to build a certain section of track

$z = .38$	$z = ./8$
Percentage = 14.80 per-cent	Percentage = 28.23 per-cent

$$28.23 - 14.80 = 13.57 \text{ percent}$$

Are there other ways to use the normal curve and z scores besides computing percentiles in frequency distributions?

Yes. The normal curve provides the fundamental basis for all of inferential statistics. But before we go any further, let's spend some time understanding what variation means when two variables are involved.

How Much Do You Remember?

New Words	New Symbols
Range	Q
Interquartile range	M.D.
Mean deviation	X^2
Variance	σ^2
Standard deviation	σ
Standard score	s^2
Normal curve	s
Standard normal distribution	z
z score	$\mathcal{N}$

Did You Ever Wonder?

1. Does the definitional formula for the standard deviation take longer to compute by hand than the computational formula? Try it both ways.
2. It is possible for the range to be larger than the variance in a distribution?
3. Can the table of Z values be translated into proportions?

Want to Know More?

1. BLALOCK, H. M. *Social Statistics.* 2nd ed. New York: McGraw-Hill Book Company, 1972. Chapter 7.
2. HAYS, W. L. *Statistics.* New York: Holt, Rinehart and Winston, Inc., 1963.
3. LEVIN, JACK. *Elementary Statistic in Social Research.* New York: Harper & Row, Publishers, 1973.
4. WILLIAMS, FREDERICK. *Reasoning with Statistics.* New York: Holt, Rinehart and Winston, Inc., 1968. Chapter 3.

Prediction

TECHNIQUES
FOR TWO VALUABLES

7

Why predict one variable from another?

Whenever we observe an event, it is our nature to try to identify its cause. This adds structure to our life and allows it to run more smoothly. If cause cannot immediately be determined, we often try to find variables which are associated with the event in order to explain it. For instance, if water drops fall on our head, we look up to see if it is raining. If we see a plane is flying extremely low, we try to determine that an airport is close by. We assume this kind of causal relationship between phenomena and are ready to anticipate it. Therefore, when new events confront us, we are able to deal with them. However, since cause cannot be observed directly, but is reasoned from the closeness and covariation of two things in time and space, we are often left with trying to identify circumstances which aid us in anticipating this covariation. Those situations in which one variable comes before others in time, and contributes to the outcome of those others, are good candidates for causal inferences. In short, whenever we use knowledge about one event in order to anticipate the outcome of another, we are utilizing the tool of **prediction**.

For each level of measurement in statistics, we have techniques for predicting one variable from another. *When our values are interval level, we utilize the concept of* **regression**. *The variable whose values* *are being predicted is called the* **dependent variable** (i.e., its outcome is "dependent" on the other variable) and can be measured along one axis of a graph. *The variable whose values we are using to predict is called the* **independent variable** (because its outcome is "independent" of the other variable) and is measured along the other axis of a graph. (See fig. 7-1.)

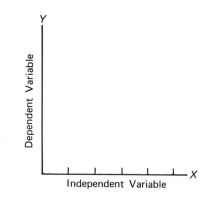

FIGURE 7-1

How many groups are involved in regression analysis?

It depends; usually the same subject is measured on both the independent and dependent variable; therefore we have only one group. For example, we observe a class of individuals and for each we might measure height and weight, I.Q. and Grade Point Average, or education and income. However, often we want to predict the value of one variable from the same variable measured in another group. For example, as mentioned in appendix I, Galton's interest was to predict the stature of sons from the stature of fathers. Here, the dependent and independent variable are the same, but the groups differ.

Both these situations occur and neither is preferred; it depends on the nature of the research. Nevertheless, the same technique for prediction is used in each.

How do we know which is the independent variable?

We have two ways of determining the independent variable. The first is to *experimentally* control the situation. If I am interested in how fast a rat will run a maze, and I think the amount of adrenalin in the blood stream has something to do with that ability, I can control the sequence and proportion of the drug, injecting differential amounts to different groups of rats in order to establish a prediction rule for maze-running speed. Thus, for example, I could predict for injections of 10 cc, 20 cc, 30 cc, etc. Clearly I have brought this independent variable to the situation by "creating" its values and therefore its effect.

A second way of determining the independent variable is to *statistically* control the data in such a way that we can observe particular levels of one variable in their natural occurrence. For instance, I may be interested in knowing whether the number of traffic tickets received is a good predictor of traffic accidents. In this example, both variables occur naturally in the situation for each case in the sample. However, we determine the number of tickets individuals have and from this develop a rule to predict the number of traffic accidents.

Both situations (the experimentally controlled or the naturally determined values) may sound vaguely alike. The key difference is that in one we control the administration of the independent variable, and in the other we must rely on its natural occurrence.

What is the concept behind the procedure
for predicting one variable from the other?

Taking the set of values for the variable $Y = 3, 4, 5, 6, 7$ and another set of values for $X = 4, 6, 8, 10, 12$, could we construct a rule for determining the values of Y from the corresponding values of X? A moment's reflection would probably yield the equation $Y = 1 + \frac{1}{2}X$. Like a recipe, this is a prescribed way to grind out the value of Y given a value of X. It is a rule for getting from one to the other value; a type of function in the sense that for each value of X, there is one and only one value of Y.

You may remember from your high school algebra class the point-slope formula for a straight line: $Y = a + bX$, a being the value on the vertical axis (ordinate) where the line crosses and b being the slope of the line; that is, for every unit change in X, there are b units change in the value of Y. (fig. 7-2). For every X-value input, you can compute a value of Y. Our example is just a special case of the formula for a straight line. In this case, the slope is $\frac{1}{2}$ and the line passes through the ordinate at $Y = 1$; hence $a = 1$.

But, suppose we introduce another set of numbers, such that $Y = 2.90, 3.95, 5.17, 6.10, 7.04$; and I asked you to again determine the recipe for finding the values of Y from the original set of X values ($X = 4, 6, 8, 10, 12$). You would be hard pressed to state on exact

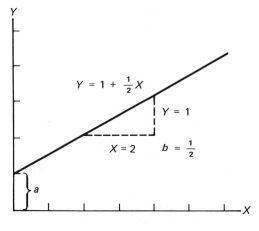

FIGURE 7-2

function rule. Nevertheless, $Y = 1 + \frac{1}{2}X$ is still a rather good approximation. The general form of this equation, $Y' = a + bX$ (where Y' indicates the predicted values of Y), can be understood as a "best guess" of the Y's if we think of these new values not as single observations but as means of separate distributions of Y values. Since we conceive of each X value as being represented by more than one Y value, we

Like a recipe, regression is a prescribed way to grind out the value of Y given a value of X

cannot call our recipe a function anymore. However, we will see that it still provides the best guess of the Y values.

In statistics, *the recipe for a straight-line prediction is called a* **linear regression** *equation.* The line it defines is a regression line. This regression line determines the best guess for values of Y given values of X. In other words, by choosing a particular value of the independent variable X, and plugging it into our equation, we can compute a predicted value of Y that will be the best guess using the *linear* regression technique.

What is the difference between using the mean of Y as the "best guess" and the regression equation as the "best guess"?

If we knew nothing of the variable X or its values, the mean of $Y(\bar{Y})$ would be the best guess of any particular Y value, merely because it would yield less error in prediction. However, if we do know something about the individual values of X (either having brought them to the situation experimentally or having statistically controlled them in their natural occurrence), we can use this information to help predict Y more accurately.

Let me elaborate. If we label the vertical axis of a graph Y, you should understand that there is a distribution of values, with a mean (indicated by a horizontal line extending to the right) labeled $\bar{Y}$. If we isolate values on the X axis, it should be evident, in light of our last example, that there are also distributions of Y values for each X. We often call these "conditional" distributions because they vary according to the value or condition of X. We can plot these values, indicating each respective mean. We improve our prediction of Y by knowing the mean or "best guess" for *each* value of X, rather than just the overall mean of $Y(\bar{Y})$.

Consider weight and height as an example. If you select a height of 5 ft 6 in., certainly I recognize that there is a distribution of weights commonly associated with this height. Therefore, my best guess would be the mean associated only with the weights of persons who are 5 ft 6 in., rather than the mean weight of all persons. The variance of the distribution of weights for people specifically 5 ft 6 in. is smaller and thus the individual mean for that value will be a better guess (i.e. yield less error) than the overall mean $\bar{Y}$. Since this is true for each successive value of X, it should be evident that the equation for a line passing closest to all these conditional values will prove the best predictor or "best guess" of the Y values. Think of the regression line Y' as a floating mean, the fit of a straight line in predicting one variable from the other (fig. 7-3).

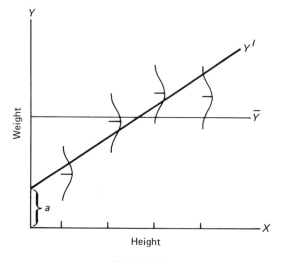

FIGURE 7-3

Why is it called linear regression?

It is linear because the regression equation is the same as the equation in our high school algebra class describing a straight line; linear, meaning straight line, is implied in the fact that all the terms in the equation are first power (i.e., X rather than X^2 or X^3). Historically, it was called regression to refer to the tendency of the estimated dependent variable to show less variation when using values of a second variable to predict (see appendix I).

What values do the symbols in the regression equation represent?

Remember we said that the regression line is the best guess in that it comes closest to all the values in the conditional distributions of Y. If it is analogous to a mean, it should have the least-squares property of a mean, which states that the sum of the squared deviations from that line are at a minimum. That is, if we took each of the distances from the individual values to the regression line and squared them, their sum would be less than the sum of squared deviations from any other straight line. Since we are predicting Y from X, a moment's reflection on the graph would indicate that we are talking about the vertical distances from the regression line.

By use of some elementary tools, we can show that finding values of a and b that minimize the distance from specific scores to the line

$Y' = a + bX$ results in two rather simple formulae:

$$b = \frac{\dfrac{\Sigma(X_i - \bar{X})(Y_i - \bar{Y})}{N}}{\dfrac{\Sigma(X_i - \bar{X})^2}{N}} = \frac{\Sigma(X_i - \bar{X})(Y_i - \bar{Y})}{\Sigma(X_i - \bar{X})^2}$$

$$a = \frac{\Sigma Y_i - b\Sigma X_i}{N} = \bar{Y} - b\bar{X}$$

For the value of a we are merely finding the Y intercept. Notice that if $b = 1$ and the means of X and Y are the same, the value of a is zero. This suggests that b "adjusts" the tilt of the line to determine where a intersects the Y axis, by specifying that proportion of $\bar{X}$ which we subtract from $\bar{Y}$. Notice also that in the formula for b the denominator is just the variance in X. The numerator is similar but rather than squaring the deviation in X, we multiply this by its corresponding deviation in Y. These paired deviations, divided by N, are called the covariance. Thus, *the regression slope of* Y *on* X (b_{YX}), *is defined as the ratio of the* **covariance** *to the variance in* X. But unlike a variance, this covariance can take on negative values; that is, if for most positive deviations in X there are corresponding negative deviations in Y, and if for most negative deviations in X there are positive deviations in Y, the resulting numerator will be negative. Therefore b can be positive or negative.

Take the example of height and weight again. If we measured these two variables, in a group of students for example, and through our formulae computed a to be 62 and b to be .85, we could predict a weight Y' for any given height. Thus, for persons 5 ft. 6 in. (65 in.), our best guess would be:

$$Y' = a + bX$$

$$Y' = 62 + .85(65)$$

$$Y' = 62 + 55 = 117 \text{ lbs}$$

Isn't there error in using regression to predict?

There can be. But just as there is a standard deviation to quantify the error in using the mean as a best guess, so there is a standard deviation of the estimate of Y, given that we use the regression line for prediction. Since it is an error in the estimation of Y, we call it the **standard error of the estimate**. Symbolically it is $s_{Y \cdot X}$, the standard deviation in Y *given* the value of X. Thus, the more widely spread the scores around the regression line, the more the error of estimation. Like any standard deviation it is defined as the square root of the variance in scores. In this case, the scores are deviations from the regression line (fig 7-4).

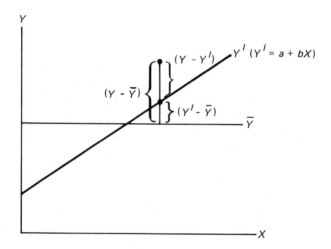

FIGURE 7-4

$$s_{Y \cdot X} = \sqrt{\frac{\Sigma(Y_i - Y')^2}{N}} = \sqrt{s_Y^2 - s_{Y'}^2}$$

The total variation in Y or $\Sigma(Y_i - \bar{Y})^2/N$ minus the reduction in variation which is accounted for by the regression equation $\Sigma(Y' - \bar{Y})^2/N$ is equal to the amount of unexplained or error variation $\Sigma(Y - Y')^2/N$. The square root of this is the standard error of the estimate. In the next chapter we will discover a rather simple way to compute this value using other measures.

Can more than one independent variable be used to predict?

    Yes. *The technique of using more than one independent variable to predict a dependent variable is called* **multiple regression.** We are testing the regression of Y on multiple variables $(X_1, X_2,$ etc.). In this case, we must estimate separate regression slopes b for each of the independent variables. For example, our equation might look like $Y = a + b_1 X_1 + b_2 X_2$ for two independent variables. The formulae for calculating these b's are analogous to the single-slope equation (for more information, see references at end of chapter). In this two-variable prediction, however, we are looking for the best regression plane rather than line, since we are dealing in two dimensions (fig. 7-5).

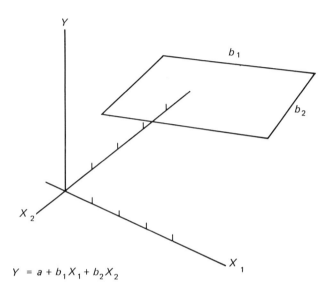

$$Y = a + b_1 X_1 + b_2 X_2$$

FIGURE 7-5

The two values of b define this plane and can be thought of as the amount of variance in Y explained independently by each of the X's. Obviously this could be extended to any number of independent variables, the assumption always being that each of these variables' effects are independent of the other and additive.

Can any variable be the dependent one in regression analysis?

Yes. By definition, the independent variable is experimentally or statistically controlled and the prediction of values are made on the dependent variable. For example, in predicting weight (Y) from height (X), we isolated values of height and identified the weight distribution of each particular height value, computing the regression of Y on $X(b_{YX})$. It should be clear that we could have isolated values of weight and looked at the conditional distributions of the heights by computing the regression of X on $Y(b_{XY})$. In doing this we are constructing a new regression line X' that comes closest to the distributions of X for each Y value. The point is that we can predict in either direction. The next chapter extends this idea of symmetry to a technique which measures the "co-relation" of two variables.

How Much Do You Remember?

New Words	*New Symbols*
Prediction	Y'
Regression	a
Dependent variable	b
Independent variable	$s_{Y \cdot X}$
Linear regression	
Standard error of the estimate	
Multiple regression	

Did You Ever Wonder?

1. How could you determine the independent variables in a survey on marijuana smoking?
2. How is a regression estimate different from an ordinary mean?
3. Is it sensible to speak of multiple regression where the predictor variables are not independent or additive?

Want to Know More?

1. BLALOCK, H. M. *Social Statistics.* 2nd ed. New York: McGraw-Hill Book Company, 1972. Chapters 17 and 18.
2. FREUND, J. E. *Modern Elementary Statistics.* 4th ed. Englewood Cliffs, N.J.: Prentice-Hall, Inc., 1973. Chapter 15.
3. KELLEY, F. J., et al. *Multiple Regression Approach.* Carbondale, Ill.: Southern Illinois University Press, 1969.
4. ROSCOE, J. T. *Fundamental Research Statistics for the Behavioral Sciences.* 2nd ed. New York: Holt, Rinehart and Winston, Inc., 1975. Chapters 14 and 42.

Correlation

REALIZING RELATIONSHIPS

8

What is correlation?

It is the "co-relation" between two variables. Since typically it is derived from the variation in two interval distributions, we often use it as a descriptive measure of the relationship between two quantitative variables—how they go together or co-vary.

Think of this analogy. You have two friends—Bill and Mary. If every available hour Bill is free, Mary also is likely to have free time and spends it with him, and every free hour Mary has, Bill is likely to be available to spend it with her, you might be tempted to say that this covariation of free time indicates they are going together. More to the point, you might conclude there is a relationship between them. Conversely, if every free minute Bill has is spent avoiding Mary and vice versa, you might say there is a negative relationship between them.

Translating this into numbers, the analogy should become clear. Assume I measure two variables, like the height of fathers and their sons. If tall fathers tend to have tall sons and short fathers tend to have short sons, I say there is a positive relationship or positive correlation between the height of fathers and sons. On the other hand, if I measure prejudice and education, and find that people with higher education are less prejudiced and people with lower education are more prejudiced, I say that there is a negative relationship or negative correlation between

You might be tempted to say this covariation in free time
indicates Bill and Mary go together

education and prejudice. Similarly, if I measure the amount of rainfall
in California and the number of babies born in New York, and find
that with an increase (or decrease) in rainfall the distribution of babies
born remains the same, I say there is no relationship or correlation.
In general, whenever an increase in the value of one variable is
accompanied by an increase (or decrease) in the value of the other variable,
we say there is a positive (or negative) correlation between the two
variables. The best way to conceive of a correlation is to think of it
as a graph of two variables in which both the abscissa and the ordinate
represent interval scales. We then plot a pair of values for each object
in the collection. Such a graph, called a **scattergram**, refers to the scatter
of pairs of values in the graph; e.g., one height for a father, a corresponding height for his son (fig. 8-1). *The **correlation coefficient** (called the
product moment correlation when the data are interval) is a means of*

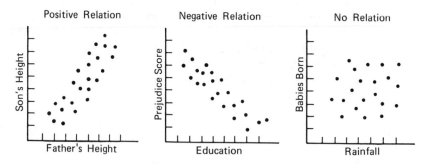

FIGURE 8-1

*describing the covariation of two interval-level variables using information
derived from the spread or variance of the scores of each.*

Is correlation different from regression?

They are both measures of two variable relations. The difference is a matter of what question is being asked, and the computation
is, therefore, slightly different. You will recall we stated that linear
regression is a description of the fit of a straight line to predict one
variable from the other. We compute this by deriving a least-squares
regression line ($Y' = a + b_{YX}X$).

Correlation, on the other hand, does not describe the fit of a straight
line in predicting one variable from another. Rather it is a measure
of the *strength* of that prediction and is symbolized by *r*. It is inversely
related to the concept of error in prediction (i.e., the standard error
of the estimate); that is, the smaller the error in prediction, the greater
the correlation; the larger the error in prediction, the smaller the
correlation (fig. 8-2). In this sense, it is a measure of how much we

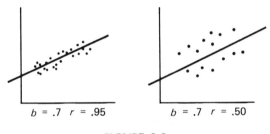

$b = .7 \quad r = .95$ $b = .7 \quad r = .50$

FIGURE 8-2

have reduced the error in predicting one variable from another using
the regression recipe. In short, regression describes the rule for prediction
of one variable from another; correlation quantifies its accuracy.

How is the correlation coefficient computed?

We compute the spread or variance of the two variables
measured. If s_X^2 is the variance in X and s_Y^2 is the variance in Y, then
$s_X^2 \cdot s_Y^2$ can be thought of as the total possible variation in both variables.
If we take the covariance, square it, and treat it as proportion of this
total variation, we have computed r^2 (remember, the numerator of the
covariance is the deviation in X times the deviation in Y). The correlation
coefficient is just the square root of this. Thus

$$r^2 = \frac{\dfrac{(\Sigma xy)^2}{N^2}}{\dfrac{\Sigma(x)^2 \cdot \Sigma(y)^2}{N^2}} = \frac{(\Sigma xy)^2}{\Sigma(x)^2 \cdot \Sigma(y)^2}$$

$$r = \sqrt{\frac{(\Sigma xy)^2}{\Sigma x^2 \cdot \Sigma y^2}} = \frac{\Sigma xy}{\sqrt{\Sigma x^2 \cdot \Sigma y^2}}$$

where $x = (X - \bar{X})$ and $y = (Y - \bar{Y})$

We have created still another new word in the vocabulary of statistics by using a combination of already familiar terms. In practice, this value can be computed from raw scores using an extension of the same computational formula that is used for the variance (see chap. 6). We often see either formula in the literature:

$$\frac{\Sigma xy}{\sqrt{\Sigma x^2 \cdot \Sigma y^2}} = \frac{\Sigma(X - \bar{X})(Y - \bar{Y})}{\sqrt{\Sigma(X - \bar{X})^2 \Sigma(Y - \bar{Y})^2}}$$

$$= \frac{N\Sigma XY - (\Sigma X)(\Sigma Y)}{\sqrt{[N\Sigma X^2 - (\Sigma X)^2][N\Sigma Y^2 - (\Sigma Y)^2]}}$$

If the data are grouped, merely replace each X or Y with the frequency times the category midpoint (i.e., $f_i m_i$).

Suppose, for example, you were employed by a federal housing agency and were interested in knowing whether there was any relationship between the number of children in a family X and the number of bedrooms in the household Y. By interviewing five families you obtained the following data.

TABLE 8-1.

	X	Y	X^2	Y^2	XY
Family 1	1	2	1	4	2
2	2	3	4	9	6
3	3	4	9	16	12
4	3	2	9	4	6
5	4	6	16	36	24

$$\Sigma X = 13 \quad \Sigma Y = 17 \quad \Sigma X^2 = 39 \quad \Sigma Y^2 = 69 \quad \Sigma XY = 50$$
$$(\Sigma X)^2 = 169 \quad (\Sigma Y)^2 = 289$$

$$r = \frac{5(50) - (13)(17)}{\sqrt{[5(39) - 169][5(69) - 289]}} = \frac{250 - 221}{\sqrt{(26)(56)}} = \frac{29}{38.2} = .76$$

Using our computational formula it is necessary to find the sums of each variable $(\Sigma X, \Sigma Y)$, the sums of the squares of each value $(\Sigma X^2, \Sigma Y^2)$, the square of the sums $[(\Sigma X)^2, (\Sigma Y)^2]$, and finally the cross product (ΣXY). Once these are determined, it is merely a matter of plugging the values into the formula as above.

For all practical purposes, this coefficient will probably be given to you via a calculator or computer. Therefore, it is much more important that you intuitively understand the concept.

How is the correlation formula interpreted?

The formula should look somewhat like that defining the slope of the regression line b. In fact, by looking at the regression of Y on X and the regression of X on $Y(b_{YX}, b_{XY})$, we can see that the correlation coefficient can be defined in terms of them:

$$b_{YX} = \frac{\Sigma xy}{\Sigma x^2} \qquad b_{XY} = \frac{\Sigma xy}{\Sigma y^2}$$

$$r^2 = b_{YX} \cdot b_{XY} \qquad r = \sqrt{b_{YX} \cdot b_{XY}}$$

In other words, the correlation coefficient is the square root of the product of the two regression slopes. Therefore, another way we can think of r is as a kind of "average" of the two linear predictions. If both the regression of Y on $X(b_{YX})$ and the regression of X on $Y(b_{XY})$ are 1.00, meaning that for any one unit change in one variable, we would predict a one unit change on the other, the correlation coefficient would = 1.00 (thus $r = 1.00$). Conversely, if both regression slopes showed inverse prediction, the correlation coefficient, and thus r, would be -1.00. The extent to which these two regression slopes are different is the extent to which the correlation coefficient will be closer to .00 (fig. 8-3). Because of the nature of this relation, we can show that

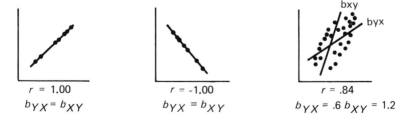

		bxy
$r = 1.00$	$r = -1.00$	$r = .84$
$b_{YX} = b_{XY}$	$b_{YX} = b_{XY}$	$b_{YX} = .6\ b_{XY} = 1.2$

FIGURE 8-3

the correlation coefficient varies between ± 1.00. A value of $+1.00$ is a perfect positive relationship; a value of .00 indicates no relationship; a value of -1.00 is a perfect negative relationship. Values in between are judged low to high depending on the size. Further notice that r^2 varies between 0 and 1.00. This has the advantage of being thought of as the **proportional reduction in error** of prediction for one variable, given knowledge of the other.

How is the correlation coefficient a measure
of the proportional reduction in error?

You will recall that given our recipe for prediction in the regression equation, the amount of error for any given estimate was called the standard error of the estimate ($s_{est\ Y}$ or $s_{Y \cdot X}$). This is analogous to the standard deviation in any distribution in the sense that it is the measure of spread or "error" around a mean. In this case, the mean is the value predicted using the regression equation. The square of this standard error of the estimate can be interpreted as that variance or error still remaining after "adjusting" our guesses of Y by computing the regression of Y on X. As stated earlier, it is the difference between the total variance around the mean of $Y(\bar{Y})$, and the variance of the regression line (Y') from the mean of Y. Thus: $s_Y^2 - s_{Y'}^2 = s_{Y \cdot X}^2$. If we express each of the terms in this equality as a proportion of the total variation, the meaning of r^2 as a proportional reduction in error should be clear. Rearranging our earlier statement we can show:

$$s_Y^2 = s_{Y'}^2 + s_{Y \cdot X}^2$$

$$\text{total} = \text{explained} + \text{unexplained}$$
$$\text{variation} \qquad \text{variation} \qquad \text{variation}$$

$$s_{Y'}^2 = s_Y^2 - s_{Y \cdot X}^2$$

$$\frac{s_{Y'}}{s_Y^2} = \frac{s_Y^2}{s_Y^2} - \frac{s_{Y \cdot X}^2}{s_Y^2}$$

$$r^2 = 1 - \frac{s_{Y \cdot X}^2}{s_Y^2}$$

Thus, r^2 is the proportion of the explained variance. It is often called the **coefficient of determination**, that is, how much knowledge of one variable helps in "determining" the values of the other. It is a proportional reduction in variance in the sense that it identifies that proportion of the total variance which has been "reduced" or explained by another variable. With this information, it is easy to show that we can define the standard error of the estimate in terms of the correlation coefficient:

$$r^2 = 1 - \frac{s^2_{Y \cdot X}}{s^2_Y} \qquad 1 - r^2 = \frac{s^2_{Y \cdot X}}{s^2_Y}$$

$$(1 - r^2)s^2_Y = s^2_{Y \cdot X} \qquad \sqrt{1 - r^2}\, s_Y = s_{Y \cdot X}$$

We often call $1 - r^2$ the **coefficient of alienation**. Remember the name by thinking that we would be alienated, or in error, a certain percentage of the time in making the least-squares estimate; $s_{Y \cdot X}$ is a measure of that error.

Since the correlation coefficient is easier to compute than the standard error of the estimate, we often compute the standard error using $1 - r^2$.

What values of r indicate a strong relationship?

Whereas the size of r^2 has a direct interpretation indicating the proportion of variance explained by using one variable to predict another, r has only a "relative" interpretation. That is, we cannot say that a correlation of .5 is twice as strong as a correlation of .25, only that it is stronger. This kind of ordinal thinking is the only meaningful way to compare the size of different r's. While there is no statistical basis for concluding what constitutes a strong relationship, tradition tells us that in social science r's ranging from .6 to .8 indicate quite strong linear relationships; from .4 to .6 moderate; and .2 to .4 low.

Does an insignificant or low correlation indicate no relationship between the variables?

No! Remember that the correlation coefficient is a measure of the strength of the ability to make *linear* predictions from one variable to another. It is quite possible that we could have a relationship like age and divorce rate where both younger and older couples experience higher rates and those in between experience lower rates (fig. 8-4). Such an association is a curvilinear relationship (i.e., the scattergram follows a curved line). A low or insignificant correlation could indicate that a rather potent relationship exists which is not linear. It is always wise, therefore, to draw a scattergram before computing the correlation

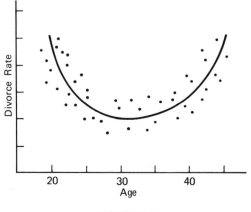

FIGURE 8-4

coefficient. It will save time and indicate the approximate form of the relationship.

Does a significantly high or perfect linear relationship indicate a causal relation?

Absolutely not! Correlation measures the strength of the linear ability to predict *either variable from the other.* A perfect correlation of 1.00 would indicate that either variable could be predicted from the other without error. Therefore it is irrelevant, statistically, to ask which causes which.

The only notion of cause available to the human senses is the idea of time and space. If one variable comes before the other in time or seems to covary in space, we may be in a better position to assert that it causes the other. However, there is nothing inherent in the correlation coefficient statistic which allows this speculation. It is purely a philosophical, not a statistical, problem.

Can the relationship between more than two variables be described using the correlation coefficient?

Yes. Two different methods can be used to describe more than two variables. The first technique, called **multiple correlation**, *measures the relationship between one variable and two or more others.* It is symbolized by *R*. The other technique, seen more often in the literature, is called **partial correlation**. *It is the relationship between two variables, holding the values of a third constant.* Partial correlation is a more effective method in social science in that it allows for the opportunity to test the effects of intervening variables by controlling the relationship of any two.

How do we use partial correlation?

Suppose you know three people: Bill, Jim, and Mary. Let's say that you observe Jim and Bill together much of the time, enough to say that there is some relationship between them. On closer inspection, however, you find that Jim and Bill are together only in the presence of Mary; that is, when Mary is not around, Bill and Jim are rarely together. It might be a logical conclusion that Jim and Bill are brought together because they are mutual friends of Mary. Without Mary present the relationship between Bill and Jim is nonexistent.

The same is true of correlations. A partial correlation measures the relationship present between two variables when the effects of a third are taken from the situation. The formula for the partial correlation is quite simple and involves only correlation coefficients:

$$r_{XY \cdot Z} = \frac{r_{XY} - (r_{XZ})(r_{YZ})}{\sqrt{[1 - (r_{XZ})^2][1 - (r_{YZ})^2]}}$$

This is read, the correlation of X and Y given (remember the dot represents the word given) variable Z. The interpretation is intuitive if we remember what we know about the correlation coefficient, and our analogy.

The relationship between X and Y given Z is equal to the proportion of total covariation between X and Y when that shared with Z is "partialed" out.

A good example might be the correlation between age X and income Y. It seems a good case to make that the older a person is, the higher the income, and the younger the less the income. A closer inspection, however, might yield different interpretations. Take the third variable education Z. From the following data we might have

$$r_{XY} - .57$$

$$r_{XZ} = .69$$

$$r_{YZ} = .72$$

$$r_{XY \cdot Z} = \frac{.57 - (.69)(.72)}{\sqrt{1 - (.69)^2} \sqrt{1 - (.72)^2}} = \frac{.07}{(.72)(.69)} = \frac{.07}{.50} = .15$$

$$r_{XY \cdot Z} = .15$$

(Remember $r_{XY} = .57$)

We can see that most of the relationship between age and income is accounted for by the amount of education one has. It is the same thing as measuring the correlation between age and income within each educational group.

Is multiple correlation used differently than partial correlation?

Partial correlation is a way of measuring the strength of the linear relationship between two variables controlling for a third, by partialing out the shared variance with that third variable. Multiple correlation, on the other hand, is a technique for measuring the strength of the linear relationship between one variable and two or more others. It is symbolized by $R_{X \cdot YZ}$ and reads the multiple correlation of X with Y and Z.

If the variables Y and Z are independent (i.e., uncorrelated) the square of this multiple correlation coefficient is merely the sum of the two correlations squared

$$R^2_{X \cdot YZ} = r^2_{XY} + r^2_{XZ}$$

If the two variables are themselves correlated, however, we must subtract out that amount of intercorrelation.

$$R^2_{X \cdot YZ} = \frac{r^2_{XY} + r^2_{XZ} - 2 r_{XY} r_{XZ} r_{YZ}}{1 - r^2_{YZ}}$$

We are saying that the amount of shared variation that X has with Y and Z together is equal to the amount it shares with each separately, minus the intercorrelation among X, Y, and Z, expressed as a proportion (i.e., divided by) of the total possible shared variation. The multiple correlation coefficient is simply the square root of this value.

Think of the age X, income Y, and education Z example once again. The squared correlation of age with income and education $R^2_{X \cdot YZ}$ can be thought of as the amount of variation that age has with income and education separately $(r^2_{XY} + r^2_{XZ})$ minus the amount of shared variation $-2 r_{XY} r_{XZ} r_{YZ}$ expressed as a proportion of the total variation possible between age and the other two variables $(1 - r^2_{YZ})$:

$$R^2_{X \cdot YZ} = \frac{.33 + .48 - 2(.57)(.69)(.72)}{1 - (.69)^2} = \frac{.81 - 2(.28)}{.52} = \frac{.25}{.52} = .49$$

$$R_{X \cdot YZ} = \sqrt{.49} = .70$$

(Remember $r_{XY} = .57$)

The squared multiple correlation coefficient $R^2_{X \cdot YZ}$ can be interpreted using the notion of proportional reduction in error. In this case we are saying how much more variance in age can be explained using both income and education ($R^2_{X \cdot YZ}$), rather than just income (r^2_{XY}).

From this brief introduction, it should be apparent that partial and multiple correlation can be extended to any number of quantitative variables. The interested student should check references at the end of the chapter for more information.

> *Does the size of r depend on the number of cases*
> *used to measure it?*

The size of r will not necessarily depend on the number of cases. However, the *significance* of the value will. That is, the larger the number of cases sampled to represent the relationship in a given population, the more confident you can be that any particular value of r is a meaningful representation of the actual relationship in the population. To answer this properly, however, requires an understanding of probability and sampling. The next section deals with questions of this sort.

How Much Do You Remember?

New Words	*New Symbols*
Scattergram	r^2_{XY}
Correlation coefficient	r_{XY}
Proportional reduction in error	$1 - r^2_{XY}$
Coefficient of determination	$r_{XY \cdot Z}$
Coefficient of alienation	$R_{X \cdot YZ}$
Multiple correlation	
Partial correlation	

Did You Ever Wonder?

1. Can the correlation coefficient be defined in terms of standard scores?
2. Is there ever a case in which correlation can be used to infer causality?
3. Is the partial or multiple correlation coefficient better for describing the relationships like that between education, income, and occupational prestige?

Want to Know More?

1. HUFF, DARRELL. *How to Lie with Statistics.* New York: W. W. Norton & Company, Inc., 1954.

2. McCall, R. B. *Fundamental Statistics for Psychology.* 2nd ed. New York: Harcourt Brace Jovanovich, Inc., 1975. Chapter 6.
3. McCollough, C., and L. Van Atta. *Introduction to Descriptive Statistics and Correlation.* New York: McGraw-Hill Book Company, 1965. Chapters 5–8.
4. Ruyan, R. P., and A. Haber. *Fundamentals of Behavioral Statistics.* 2nd ed. Reading, Mass.: Addison-Wesley Publishing Co., 1971.

Inferences

III

Probability

THE ART AND SCIENCE
OF MEASURING CHANCE

9

What is probability?

Put most simply, a probability is a proportion, but a special type of proportion. Just as in descriptive statistics we talk of a proportion as the frequency in any given subset divided by the total N, so *we can define* **probability** *as the number of favorable events divided by the total number of events.* To be more specific, a probability is the proportion of times an event will occur given a large number of chances. Sometimes we like to call this large number of chances the "long run." Therefore, a probability is the proportion of favorable events occurring in the long run. No one is really sure just how long this is, so to get around it mathematicians have invented a term called the *limit.* In our case the limit is the value our proportion of successes approaches as the number of trials goes to infinity (i.e., gets larger and larger):

$$P(X) = \frac{f}{N} \quad \text{as} \quad N \to \infty$$

You can best understand this if you think of a penny. If I ask you what the probability is of getting a head if you flip the penny just once, a moment's reflection would tell you that it is either 0 or 1; yet, I'm sure you would answer $\frac{1}{2}$, which is, of course, impossible for one flip. But our mind tends to think in terms of what would happen over a very large number of trials.

So, even though we roll the die (one of a pair of dice) only once, we know the probability of getting any one number is $1/6$. Even though we draw a card only once, the probability of getting an ace of spades is $1/52$. And while you may only have one baby, the probability of getting a boy is roughly $1/2$. We get away with this wizardry because we can identify the total number of outcomes, and can enumerate the number of favorable events. This means we can logically determine the probability beforehand, or in other words compute the *a priori* probability.

*Why do mathematicians refer to cards, dice, and coins
when explaining probability?*

The science of probability started in the seventeenth century when gamblers sought the help of mathematicians to determine how best to win. It is no wonder, then, that most of the early works of probability dealt with game of chance.

You have noticed that almost always examples of cards and dice will come up in statistics classes. The reason is largely because mathematicians such as Pascal, de Fermat, De Moivre, and Bernoulli, were fascinated with these games. But more important, we use these games as teaching devices because they offer the most clear-cut examples of deriving mathematical probabilities, rather than the alternative of trying to determine empirical probabilities.

The science of probability started in the seventeenth century when gamblers sought the help of mathematicians to determine how best to win

*Why distinguish between mathematical and empirical
probability?*

The word probability comes in many disguises. Often the
terms "odds," "chances," or "law of averages" are substituted, render-
ing the concept more easily identifiable. Other times the concept of
probability masks itself and is more difficult to recognize. For example,
the daily sunrise and the ultimate death of a person would not ordinarily
be thought of as probabilities. They are "certainties." A probability,
on the other hand, seems to connote more doubtful matters. But
remember, a probability is a proportion—the proportion of times that
an outcome will occur. And just like a proportion, its lowest value
is zero (you can never have less than none of the successes) and the
highest value is one (you can never succeed more than all the times).
If we think of it this way, the mystery of the sunrise as a probability
disappears. What we call certainty is just a special case in which the
probability is one.

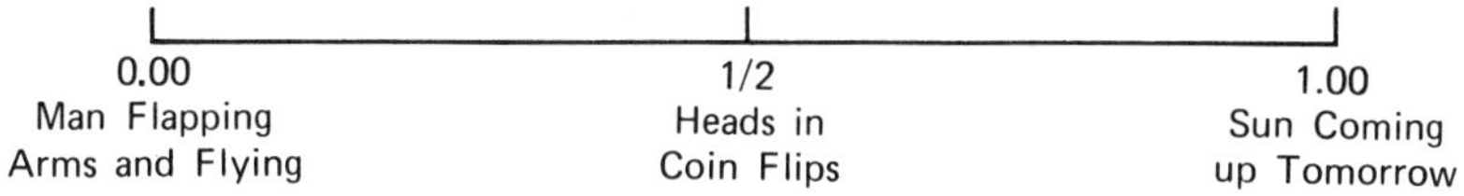

0.00	1/2	1.00
Man Flapping	Heads in	Sun Coming
Arms and Flying	Coin Flips	up Tomorrow

FIGURE 9-1

Another area seemingly divorced from probability is that of fantastic,
unlikely events that have never occurred in our experience. We don't
consider it a probability statement when we think of a man flapping
his arms to fly, or someone swimming the Atlantic Ocean. Yet isn't
this just a special case of the probability being zero? (See fig. 9-1.)

How do we resolve this apparent dilemma? If events in the real
world either happen or don't happen, how can we arrive at a proportion
that describes the likelihood of that outcome occurring? The answer,
in part, revolves around an important distinction that scientists and
mathematicians often make between **mathematical**, or a priori, probability
(logically determined before it happens) and **empirical**, or a posteriori,
probability (known after it happens). Mathematical probability rests on
the assumption that we know the total possible outcomes and can assign
probabilities to them. Thus we know that there are two sides to a coin,
and fifty-two cards in a deck, and six sides to a die. We assume each
outcome is equally likely to occur and can specify, therefore, the
probability of each. (We must, in good conscience, discount the possibility
of the coin landing on its edge or the die falling down the gutter.)

In other situations we can't always logically determine what the probability of given outcomes are. To find out, we dutifully watch and count events as we run "experiments." By observing a number of trials of the phenomenon we are interested in, recording the frequency of outcome for each alternative, we can then derive an empirical probability, which is defined as the frequency of a particular observed outcome divided by the total number of trials. This relative frequency is the empirical or a posteriori probability.

In determining empirical probability, however, we must make sure we have enough trials. I could hardly predict rain in a given area if I had information about only one other day's weather there. I would know very little about the traffic patterns in a city if I observed only during one rush hour. And obviously, I would know very little about a biased coin if I flipped it only once. The point here is that we need a suffient number of trials to determine the empirical probabilities. Yet even a very large number of observations may be insufficient to produce exact probability estimates. We can, however, state how closely an empirical probability estimate approximates a theoretically predetermined probability. Such precision, we will see, depends on the number of observations (N) in our empirical "experiment."

How are mathematical probabilities calculated?

First we must enumerate all the possible outcomes. Second, we identify an outcome whose probability we are interested in. And finally, we divide the particular outcome by the total number of outcomes. For example, the probability of a 6 coming up in one roll of a die is $1/6$. The problem is that it becomes tedious to count the total when the number of possible outcomes is large. Fortunately, there is another way to accomplish the same goal. It involves counting techniques known as permutations and combinations.

What are permutations?

Suppose you have to take your lunch to school every day, and while sitting at the kitchen table in the morning, you prop your eyes open enough to find that you have made three different kinds of sandwiches, and have two different kinds of fruit. If you take only one sandwich and one piece of fruit, how many different lunches could you have? You're right, six; that is, any piece of fruit with any sandwich. In general, we will have as many distinct arrangements as there are in the product of numbers in each category.

Now what if we have three students in a classroom and we want to know how many different arrangements we could make for them to leave the room one at a time. That is, how many different *orders* would we have for their leaving the room? Obviously, any one of the three could leave first, any one of the two remaining could leave second, and the only one left would leave last. We have $3 \cdot 2 \cdot 1 = 6$ ways these people could leave. In general we state that with N objects, we can order them in $N!$ distinct ways. This is another new symbol in our language of statistics. It stands for **factorial** and means that we multiply $N(N - 1)(N - 2)(N - 3) \ldots (1)$. Thus, in our example above we have $3!$ ($3 \cdot 2 \cdot 1 = 6$) ways of the students leaving the room. Actually the symbol for factorial ! is a result of a printer's request in the nineteenth century for a simpler way to typeset the old symbol which was ⌐ as in ⌐^{3} (even printers help to define this language of statistical symbols!).

Now let's take this one more step and assume that we aren't interested in arranging all the N objects in order. Assume we want to know the order of only, say, r of them, as in this example: You get to the race track and find that you want to bet the exacta in the eighth race (an exacta is betting on the first- and second-place finishers in their order of finish). Suppose there are five horses in the race ($N = 5$). How many different exactas can you bet on? In other words, in how many different orders can these five horses finish 1-2? If any one of the five finishes first, then any one of the other four can finish second. Thus we have $5 \cdot 4 = 20$ possible different orders to bet on. Notice we aren't using all of the $5!$, but only the first two terms: $5 \cdot 4$.

Questions of this type, in which we are interested in order are called **permutations** and are written $P(N,r)$. That is, how many permutations or orders of N things can we get taking them r at a time. We know that the special case of the permutation in which we are ordering all N things is $P(N,N) = N!$, as in our first example. When we order less than all N things, we are essentially canceling out the last $(N - r)$ terms of the factorial and so we may write it as follows:

$$P(N,r) = \frac{N(N - 1) \ldots (N - r + 1)(N - r)(N - r - 1) \ldots 1}{(N - r)(N - r - 1) \ldots 1}$$

In our horse race we have

$$P(5,2) = \frac{5 \cdot 4 \cdot 3 \cdot 2 \cdot 1}{3 \cdot 2 \cdot 1} = 20$$

In canceling the last $N - r$ terms, we are essentially dividing the permutation $N!$ by $(N - r)!$ Thus, we can rewrite

$$P(N,r) = \frac{N!}{(N - r)!}$$

Are combinations and permutations the same thing?

No! There are fewer combinations than permutations. Suppose you are slaving in your kitchen one sweltering summer night attempting to make a tasty spaghetti sauce. You have five spices in front of you, but you want to use only two of them. We know that there are $5!/3! = 20$ different orders that you could use to put the spices into the bowl. But like any impatient spaghetti eater, you are not interested in the order in which the ingredients are added, only how it tastes once they are combined. You hastily grab two spices at random and shake them briskly into the bowl. How many different potential tastes could there be?

You are not interested in the order in which the ingredients were added, only how they taste once they are combined

When we are interested in arrangements of this type without regard to order, we are dealing with **combinations**. Combinations are symbolized $C(N,r)$. That is, how many different combinations of N things taken r at a time would you have? Since we are not concerned about the order of these r objects, it should be clear that we have $r!$ (the number of ways to order r objects) more permutations than combinations for any given N. That is,

$$C(N,r)\, r! = P(N,r) \quad \text{or} \quad C(N,r) = \frac{P(N,r)}{r!}$$

But since

$$P(N,r) = \frac{N!}{(N-r)!}$$

we can write the number of combinations as

$$C(N,r) = \frac{N!}{r!(N-r)!}$$

in our spaghetti sauce example, the number of combinations is $C(5,2) =$

$$\frac{5!}{2!(5-2)!} = \frac{5!}{2!3!} = \frac{5 \cdot 4 \cdot 3 \cdot 2 \cdot 1}{2 \cdot 1 \cdot 3 \cdot 2 \cdot 1} = 10$$

This is considerably less than the 20 orders we could put them in.

The idea of a probability should become more intuitive, now, when you think of it as a ratio of combinations. For example, the probability of drawing a flush (all cards from the same suit) in a poker hand is the number of ways of drawing 5 of one suit from the total number of that suit (13) times 4 (the number of suits in a deck) divided by the total number of ways of drawing any 5 cards from a deck of 52. Symbolically it is

$$\frac{C(13,5) \times 4}{C(52,5)} = \frac{5,148}{2,598,960} = .002$$

What properties do probabilities have?

There are four. The first two define a probability distribution; the last two are rules for computing probabilities. Because a probability is a proportion it must be non-negative in value and no greater than 1. That is,

1. $$0 \leq P(X) \leq 1$$

Further, since all the outcomes of an experiment must exhaust the possibilities, all the probabilities in a distribution must sum to 1. That is,

2.
$$\sum_{i=1}^{N} P(X_i) = 1$$

This is only reasonable in that if something is going to occur, e.g., like your buying a particular car, then the probability of buying each alternative must add up to the certainty, i.e., $P(x) = 1$, that you are going to buy *a* car.

There are two other important properties. One has to do with the probability of one event *or* another occurring, and the other has to do with the probability of one event *and* another occurring. The first is called the addition rule of probability. Symbolically it is written as

3.
$$P(A \cup B) = P(A) + P(B) - P(A \cap B)$$

First, notice the funny little notation on the left side of the equation. That's right, the one that looks like an upside-down horseshoe. This is the symbol for the word *union*. This means that we want to know the probability of the outcomes *A or B* happening. If you get stuck remembering this indicates addition, just recall that the union of holy matrimony is the process of *adding* two people together. So, if we want to know the probability of two outcomes, like drawing an ace *or* spade, add the probability of drawing an ace ($4/52$) to the probability of drawing a spade ($1/4$), and subtract off the amount that they overlap (I do this since I have counted the ace of spades twice, once for aces and once for spades; thus I need to subtract one of them). Symbolically, $P(A \cup S) = 1/13 + 1/4 - 1/52 = 4/13$.

This brings us to a notation of the second part of our formula. The symbol that looks like a horseshoe means *intersection;* that is, the probability of both events occurring simultaneously. The key word is *and:* $P(A$ and $B)$ occurring. If this one is tough to remember, just think of the intersection of two streets which is the corner of such and such *and* so and so. In the card example, the ace and the spade can occur simultaneously, so we subtract it off once after counting it twice. If the outcomes are all mutually exclusive, like the probability of getting a spade or a club, then obviously there is no overlap and the last term drops out:

$$P(A \cup B) = P(A) + P(B)$$

This is easy to keep straight if we remember the analogy of a barrel of fruit. I have five apples and five oranges and five bananas. The probability of drawing an apple or an orange is $P(A \cup 0) = 1/3 + 1/3 = 2/3$. There is no overlap unless, of course, you are making fruit salad. If you are wondering how we go about computing the intersection of two events, then you will no doubt want to pay attention to the next property of probability. This is called the rule of multiplication and is defined as:

4. $P(A \cap B) = P(A)\,P(B|A) = P(B)\,P(A|B)$

This is read: the probability of *A* and *B* occurring is equal to the probability of *A* times the probability of *B* *given* that *A* has already occurred. We have introduced another new word into our vocabulary here so let me explain. *The probability of an event given that another has occurred is called the* **conditional probability**. It means that the probability is subject to a condition; that one event occurring is affected by another event having already happened. The probability of your living to be 70 years old is greatly enhanced given that you have already lived 20 years. The probability of your graduating from college is greatly increased given that you passed your first year of classes. The probability of drawing an ace is increased given that you did not get one in the first draw. However, the probability of getting a head on one flip of the coin does not affect the probability of getting one on the next flip. This means outcomes of a coin toss are independent; in the other examples, the outcomes are not. If alternative outcomes are independent, the formula for intersection reduces to

$$P(A \cap B) = P(A)\,P(B)$$

When events are not independent, we must compute or be given the conditional probabilities. Events like tossing dice, flipping coins, turning a roulette wheel, all have fixed numbers of total outcomes that don't change from trial to trial. Outcomes are independent. With a game like cards, however, the total number of alternative outcomes is not fixed and may change from trial to trial as the cards are dealt from the deck. Therefore, the outcomes are dependent. The probability of drawing exactly two aces in two draws without replacing the first card is $(4/52)(3/51)$. The probability of drawing a second ace is going to be altered given that I have one in the first draw. Notice that if I replace the first card, it can't have an effect on the second draw and the answer is $(4/52)(4/52)$. Thus, drawing with replacement assures independence while to sample (draw) without replacement means we must compute the conditional probability.

Are mutually exclusive *and* independent *probabilities the same concepts?*

Do not be fooled into thinking that these are the same. They are not. Mutually exclusive means that the two events cannot occur simultaneously. Independent means that the outcome of one event does not affect the outcome of another. To make one statement does not imply anything about the other. Events may be mutually exclusive (or not) in a single trial. But in order to establish independence, we must have information about more than one trial to determine if one outcome affects the other. You should see this if we combine the addition and multiplication formula:

$$P(A \cup B) = P(A) + P(B) - P(A \cap B)$$

but

$$P(A \cap B) = P(A)P(B|A)$$

therefore

$$P(A \cup B) = P(A) + P(B) - P(A)P(B|A)$$

if independent but not mutually exclusive, the formula reduces to

$$P(A \cup B) = P(A) + P(B) - P(A)P(B)$$

but if they are mutually exclusive, the last term drops out altogether.

Think of the cards. If I draw a single card out of a deck of 52, I can calculate the probability of two mutually exclusive events (like drawing a spade or a club; that is, a black card) or I can calculate the probability of two events which aren't mutually exclusive (drawing an ace or a spade). But I can't determine anything about the independence of the event unless I can determine the conditional probability. Since there is no conditional probability with only one trial, I must have another trial to calculate this value. I cannot determine the probability of drawing a club and a spade if I am only drawing once.

> *What is the addition and multiplication rule if we have more than two probabilities?*

Looking at these two rules of probability, taking the case of the union of several outcomes which are mutually exclusive and the intersection of several outcomes where they are independent, we can extend the definitions to more than two outcomes. Symbolically, the addition rule becomes

$$P(A \cup B \cup C ...) = P(A) + P(B) + P(C)$$

The union of more than two outcomes is merely the sum of their respective probabilities if they are mutually exclusive. The multiplication rule becomes

$$P(A \cap B \cap C \ldots) = P(A)\,P(B)\,P(C)$$

The intersection of more than two outcomes is merely the product of their respective probabilities if they are independent.

We will now take our knowledge of counting and the properties of probabilities and apply them to probability or sampling distributions, particularly two called the binomial and the normal distribution.

How Much Do You Remember?

New Words	*New Symbols*
Probability	∞
Limit	$P(X)$
Mathematical probability	!
Empirical probability	$P(N,r)$
Factorial	$C(N,r)$
Permutations	$\cup$
Combinations	$\cap$
Union	
Intersection	
Conditional probability	
Mutually exclusive	
Independent	

Did You Ever Wonder?

1. What is the probability of being dealt 21 (an ace, combined with either a 10 or a face card) from a fresh deck of 52 cards?
2. Why, when three dice are rolled, are there 27 ways to get a total of 10 while only 25 ways to get a total of 9, out of 216 possible results?
3. Is a three-person jury in which one man flips a coin for his vote more accurate in their decision than a one-man jury?

Want to Know More?

1. BOEHM, G. A. "The Science of Being Almost Certain," *Fortune* 69, no. 2 (1964): 104–107, 142, 144, 146, 148.
2. BLALOCK, H. M. *Social Statistics*. New York: McGraw-Hill Book Company, 1972. Chapter 9.
3. DRAPER, N. R., and W. E. LAURENCE. *Probability, an Introductory Course*. Chicago: Markham Publishing Company, 1970.
4. MOSTELLER, FREDRICK. *50 Challenging Problems in Probability*. Reading, Mass.: Addison-Wesley Publishing Co., Inc., 1965.

Sampling distributions and the normal curve

THE THEORETICAL THEOREM

10

Why is a sampling distribution referred to as a probability distribution?

We will soon see, in a roundabout way, that the rules of probability are the foundations for defining two special sampling distributions: the binomial and normal. But before going into that, let's lay the groundwork by explaining why a sampling distribution is a probability distribution.

We have said that in order to qualify as a probability distribution, the probabilities must be non-negative and sum to 1 (see chap. 9). Since we are defining these probabilities as the proportion of success in the "long run," it should be easy to see that a probability distribution is the distribution of outcomes which we would expect to find if we repeated our observations over and over. Such repeated observations imply that we are "sampling" randomly (i.e. looking at a randomly selected subset of outcomes) and that each sample is **independent** of the next.

Does the concept of randomness assure independence in sampling?

The concept of **randomness** has essentially two properties. First in order for a sample to be random, each case has to have an equally likely chance of being drawn. And second, each *combination* of cases must have the same probability of being chosen. In order for sampling to be independent, both these properties must be met for each draw. In simple random sampling we can assure independence by returning to the same number of total outcomes each time we select a case. With a coin this is always true; with a deck of cards, this would mean replacing the card after each draw.

If we choose outcomes without replacing them, each case does not have an equally likely chance of being chosen, therefore violating the independence rule. However, we often get around this. That is, sampling without replacement will have little effect if the number of total outcomes is very large relative to the number in our sample. This is often the case when social scientists survey large populations. Because the change in probability is so slight, relative to the number of cases chosen, random selection, and therefore independence, is assumed.

Since we are using the concept of "long run" to characterize theoretical sampling distributions, we often refer to their means as **expected values.** This should not be too surprising since we are really saying that we know what to expect our average value to be in the long run. The symbol for this is $E(X)$; that is, the expected value or average in the "long run." In a normal sampling distribution this is equal to the mean of the population μ. However, in order to understand this connection between $E(X)$ and μ, it is necessary to introduce the distribution of two outcomes, called the binomial sampling distribution.

Why is it called a binomial sampling distribution?

It is called a binomial sampling distribution because it is a probability distribution for two possible outcomes. Suppose we take a typical classroom in which half the students are boys and half girls. Our first property of probability distributions states that the probability of picking an individual of a given sex at random is going to be positive, and indeed it is $P(B) = P(G) = \frac{1}{2}$. Our second property tells us that the sum of the alternative outcomes must equal 1. It does $P(B) + P(G) = \frac{1}{2} + \frac{1}{2} = 1$. Often when we have two alternatives, we call the probability of a success (what we are interested in) P and the probability of failure (the other event) $1 - P$ or Q. (Actually, the tradition of calling them "success" or "failure" has clung tenaciously since the seventeenth century, when probabilities were computed on winning strategies in flipping coins for wagers. In their case, it was an apt term, indicating success or failure in winning money.)

Now, consider the following experiment. Suppose I pick the names of class members out of a hat and identify them as either boys or girls, replacing the choices after each draw. For the sake of this example I will do this two times. What is the probability that both draws are boys? The multiplication rule of probability tells us that it will be $1/2$ times $1/2 = 1/4$ (because they are independent draws). Likewise, the probability for girls will be $1/2 \cdot 1/2 = 1/4$. But does that exhaust the possibilities? Of course not; we could also get a girl and then a boy or a boy and then a girl. In fact, the addition rule tells us we have

$$BB = 1/2 \cdot 1/2 = 1/4$$
$$BG = 1/2 \cdot 1/2 = 1/4$$
$$GB = 1/2 \cdot 1/2 = 1/4$$
$$GG = 1/2 \cdot 1/2 = 1/4$$
$$\overline{1}$$

If we want to write this in considerable shorthand we could do it. Show the probability of drawing a girl as P and a boy as $1 - P$ or Q. Then P and Q for the two draws enumerate the entire probability distribution. Symbolically it is $(P + Q)^2$. We should recognize this again as our old friend from high school algebra known as the binomial expansion (see chap. 6 and appendix III). The name comes from two (bi) names (nomial) and we are expanding it. If we square these terms we get $(P + Q)^2 = P^2 + 2PQ + Q^2$. Substituting our values for P and Q, we find the exact distribution of probabilities as before, namely,

$$1/4 + 2(1/4) + 1/4 = 1$$

Now what happens if we choose three names? The formula would then be the sum of the probabilities cubed and would describe the entire distribution for three trials. Symbolically we have $(P + Q)^3$. This looks considerably harder to compute, but it really isn't if you think about it intuitively. For example, if I pick names three times and enumerate the number of ways to get various permutations of boys and girls I would have

$$BBB = 1/2 \cdot 1/2 \cdot 1/2 \quad \Big\} \quad 1/8$$
$$BBG = 1/2 \cdot 1/2 \cdot 1/2$$
$$BGB = 1/2 \cdot 1/2 \cdot 1/2 \quad \Bigg\} \quad 3/8$$
$$GBB = 1/2 \cdot 1/2 \cdot 1/2$$

$$GGB = \tfrac{1}{2} \cdot \tfrac{1}{2} \cdot \tfrac{1}{2} \;\Big\rbrace$$
$$GBG = \tfrac{1}{2} \cdot \tfrac{1}{2} \cdot \tfrac{1}{2} \;\Big\rbrace \quad \tfrac{3}{8}$$
$$BGG = \tfrac{1}{2} \cdot \tfrac{1}{2} \cdot \tfrac{1}{2} \;\Big\rbrace$$
$$GGG = \tfrac{1}{2} \cdot \tfrac{1}{2} \cdot \tfrac{1}{2} \qquad \tfrac{1}{8}$$
$$\overline{1}$$

Therefore, if we expand the expression $(P + Q)^3$ we would have $P^3 + 3P^2Q + 3PQ^2 + Q^3$. *In general, the distribution of probabilities derived from $(P + Q)^N$ is called a* **binomial distribution**.

Notice that the coefficients of the two-trial case were 1 2 1. In the three-trial case they are 1 3 3 1. If we keep on increasing the number of trials (and hence the power of the binomial), we find the coefficients of the terms in those expansions build on each other like a triangle. In fact, it is called *Pascal's triangle* after the famous eighteenth-century mathematician. The triangle looks something like this:

			1				$(P + Q)^0$
		1		1			$(P + Q)^1$
	1		2		1		$(P + Q)^2$
1		3		3		1	$(P + Q)^3$
1	4		6		4	1	$(P + Q)^4$
1	5	10		10	5	1	$(P + Q)^5$

This crazy pile of numbers can by derived very easily by remembering that any number is merely the sum of the two numbers above it. What it tells us is the coefficients of the binomial $(P + Q)$ raised to any particular power. These coefficients represent a given number of successes. For example $(P + Q)^4$ means that 1 4 6 4 1 are the number of ways we could get $4P$'s, $3P$'s, $2P$'s, $1P$, and $0P$'s respectively. If pyramids give you the jitters, there is another way to remember the same thing using combinations.

How do combinations help count in a binomial sampling distribution?

Take the combination $C(4,3)$. This is equal to $4!/3!1!$. Think of this as saying, "Given 4 things, how many collections have we where there are 3 of one and 1 of the other?" Remember our example in the classroom where we draw names indicating either boy or girl, replacing the name. If I ask how many different collections of 3 girls and 1 boy can I form, you would tell me

$$4!/3!1! = \frac{4 \cdot 3 \cdot 2 \cdot 1}{3 \cdot 2 \cdot 1 \cdot 1} = 4.$$

Now go back to Pascal's triangle. You will find this is exactly the same number indicated as the coefficient of the second term of the binomial expansion $(P + Q)^4$. Because of this relation with combinations, the formula defining the general binomial expansion $(P + Q)^N$ is expressed as $C(N,r) P^r Q^{N-r}$. In the binomial distribution we usually write $C(N,r)$ as $\binom{N}{r}$. Hence the formula enumerating all the probabilities in the binomial sampling distribution (which of course must sum to 1) is

$$(P + Q)^N = \sum_{r=0}^{n} \binom{N}{r} P^r Q^{N-r} = 1$$

This collection of symbols, indicating a new word in the vocabulary of statistics, is rather large, so reflect on it a moment. If we want to know the probability of any one of the possible outcomes, such as the probability of drawing 3 girls and one boy in 4 draws, we know the value would be the result of the following term in the binomial sampling distribution of probabilities:

$$\underbrace{\underbrace{P \cdot P \cdot P}_{r} \cdot \underbrace{Q}_{n-r}}_{N} \times \binom{4}{3} \text{ ways it could happen}$$

$$G \cdot G \cdot G \cdot B = 1/2 \cdot 1/2 \cdot 1/2 \cdot 1/2 = 1/16 \times 4 \text{ ways} = 4/16 \quad \text{or } .25$$

$$\text{where} \quad N = 4 \quad r = 3 \quad P = .5 \quad Q = .5$$

One more example. What would be the probability of the Los Angeles Dodgers winning a three-game series if the probability of winning each game is .6 (i.e., $P = .6$)? Since N is 3 and P is .6, this is all we need to know to specify the entire probability distribution. But specially we want to know the probability of winning all three; therefore, we calculate

$$N = 3 \quad r = 3 \quad P = .6 \quad Q = .4$$

$$\binom{3}{3} (.6)^3 (.4)^0 = 1 \text{ way} \times (.216) \times 1 = .216$$

Again, deriving this value from a sampling distribution, we are saying that given a very large number of samples of three games we would expect the Dodgers to win all three of them .216 of the time. (assuming that for each game the probability of winning is .6).

Is the binomial the same kind of distribution
as the normal distribution?

The binomial expansion $(P + Q)^N$ led directly to De Moivre's first definition of the **normal distribution** or curve of errors (see appendix I). The definition has since been extended from the two outcome to the continuous case; however, the binomial distribution helps to understand how the process started. So, gather up your new knowledge of the z-score, probability, and the binomial distribution and let's take a look at the formula that defines the probability of a given outcome for the normal curve (fig. 10-1).

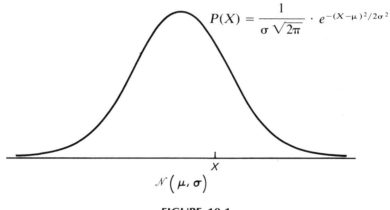

$$P(X) = \frac{1}{\sigma \sqrt{2\pi}} \cdot e^{-(X-\mu)^2/2\sigma^2}$$

$$\mathcal{N}(\mu, \sigma)$$

FIGURE 10-1

Looks pretty complicated? I would wager that the reasons for that are twofold: there are several new symbols you have never before encountered, and they are put together in a new way. But let's jump into it anyway. First a little house cleaning. If the equation is restated in standardized units we don't have to bother with raw scores. The mean will be zero and the standard deviation one (see chapter 6). If you look at the exponent of e, you will see this simplifies things tremendously. In fact, the formula reduces to

$$P(X) = \frac{1}{\sqrt{2\pi}} \cdot e^{-z^2/2}$$

Looks better already, no? Anyway, let's take each term: first $e^{-z^2/2}$. The symbol e stands for the sum of a specific series of terms. It is the mathematician's way of summing up an infinitely large number of terms in a particular expansion. In this case, the expansion is analogous to the terms in the binomial distribution. If we have $(P + Q)^N$ and take a very large value for N, then we can describe the middle term of this expansion as it relates to the sum of all the terms. This relation is reflected in $1/\sqrt{2\pi}$, the first half of the formula for the normal curve. The value of the other terms as they relate to the sum of the terms is described by $e^{-z^2/2}$, the second half of the formula for the normal curve. Therefore, what we are really saying is that if we base our binomial expansion on an increasingly large number of sample observations, the shape of the sampling distribution of successes will be approximately normal (fig. 10-2). Without looking at any more

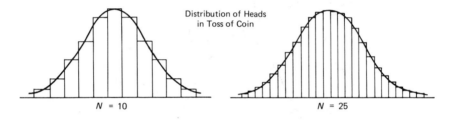

Distribution of Heads
in Toss of Coin

$N = 10$ $N = 25$

FIGURE 10-2

formulae, let's interpret values for the normal curve. The formula implies that if we want to know what value of X has the highest frequency, we must find an X which will make the exponent $z^2/2$ the smallest. Since $z = (X - \mu)/\sigma$, a value of X equal to the mean will give us an exponent of e^{-0} or $1/e^0$. Either way, the value of e is 1 since anything to the 0th power is 1. This means the entire term becomes 1 and the highest frequency is $1/\sqrt{2\pi}$ where $x = \mu$. Any other exponent will yield proportionately smaller frequency as $e^{-z^2/2}$ will become smaller.

Remember, this is in terms of standard scores. If we wanted to translate it back into raw scores, we would have to multiply by some notion of raw score units. It is more important that we keep our formula in terms of standard units, however. This is profitable in that we can view the total area under the normal curve (i.e., frequency) as one unit. This qualifies it as a probability distribution since the values sum to 1, and means that any part of the cumulative frequency is really a proportion, which after all is what a probability is!

All this tells us three things. First, we can look at the normal curve as a probability distribution. This probability distribution has one hump with the highest proportion of frequency at the mean. Second, the fact that we are dealing with a squared exponent tells us that the value of the proportion of frequencies is the same on either side of the mean (i.e., it's symmetric). Last, even though the value of the negative exponent of e can get infinitely large, which means the proportion of frequency gets infinitely small as we move out from the mean, it can never reach zero. Therefore, the tails of the normal distribution come extremely close to the horizontal axis, but never touch.

How can a frequency polygon be such a perfectly symmetric normal curve?

It can't really. It merely points to the fact that empirically if we could sample a large number of cases an infinite number of times, the resulting distribution of frequencies would look approximately normal. The normal curve is, therefore, a theoretical distribution of mathematical probabilities.

How does the normal distribution serve as a sampling distribution in statistical inference? ˙

As we have mentioned, the normal distribution, like the binomial, is based on an infinitely large sampling of independent observations. Now consider the case in which sets of observations are sampled. Each set has a mean and standard deviation for the values in it. As it turns out the distribution of these means computed from sets of observations will also form a normal distribution. The theorem that permits us to make that statement is called the Central Limit Theorem and is among the most important in all of statistics. It allows us to use information about any given sample to infer or estimate something about the population.

What is the central limit theorem?

When we introduced the normal probability distribution, we said that it had its beginnings with the binomial, or two-outcome, case. Early in the 1900's, it became apparent that this definition could be expanded to include repeated samples of continuous variables. The resulting theorem which serves to explain this is called the Central Limit Theorem. Briefly stated, the **Central Limit Theorem** says: *If we take an infinite number of random samples of size N from a population with mean μ and standard deviation σ, the distribution of the means of those samples will be approximately normal, with a mean of $\mu_{\bar{x}}$ and a standard deviation of $\sigma/\sqrt{N}$.* This theorem is extremely important in that regardless of the shape of the population distribution, repeated random sampling yields a sampling distribution of means which is normal.

This means we now have three different kinds of continuous distributions, each with a mean and a standard deviation (table 10-1).

TABLE 10-1

Type of Distribution	Mean	Standard Deviation
1. Population distribution	μ	σ
2. Sample distribution	$\bar{X}$	s
3. Sampling distribution	$\mu_{\bar{x}}$	$\sigma_{\bar{x}} = \dfrac{\sigma}{\sqrt{N}}$

Notice that the difference between the population standard deviation and the sampling distribution standard deviation is division by $\sqrt{N}$.

Why is the standard deviation divided by the $\sqrt{N}$ when computing the standard error of the mean?

When computing the standard deviation of the sampling distribution of the mean we must consider the size of the sample. Obviously, if the standard deviation is $\sigma/\sqrt{N}$ the variance of that error must be σ^2/N. We can see that as the size of our samples increases (i.e. N gets larger), the amount of error will decrease, approaching zero as a limit. This should be apparent if we observe that when the sample size is the same as the number in our population, the mean of the sample would *be* the mean of the population and the error of that sample mean in estimating μ would be zero. Any sample N less than the size of the population will produce a sample mean which contributes to the error in estimating the population mean μ. This error is defined by the variance in the population divided by the size of the sample. Because this variance σ^2/N is an index of error, we call its square root the **standard error of the mean** defined as $\sigma/\sqrt{N}$.

Can the mean of a sample be used as an accurate estimate of the population mean?

A great deal of the time our sample mean is a good estimate of the sampling distribution mean, and therefore a good estimate of the population mean. In general, using a sample to infer properties of a population is commonly known as statistical inference. Before discussing the actual process, however, we will state some fundamental rules in the next chapter about making inferences; that is, the nature of hypothesis testing.

How Much Do You Remember?

New Words	*New Symbols*
Independent	$E(X)$
Expected value	$\binom{N}{r}$
Randomness	
	$\mathcal{N}(\mu,\sigma)$
Binomial sampling distribution	$\mu_{\bar{x}}$
Normal sampling distribution	$\sigma_{\bar{x}}$
Central limit theorem	$S_{\bar{x}}$
Standard error of the mean	

Did You Even Wonder?

1. Why do researchers, who sample large cross sections of the nation's population, without replacement, still consider this an "independent" sampling?
2. Why is a binomial sampling distribution symmetric when $P = .5$?
3. Why should a sampling distribution be normal in shape even if the population distribution isn't?

What to Know More?

1. Deming, W. E. *Some Theory of Sampling.* New York: Wiley, 1950.
2. Hays, W. L. *Statistics.* New York: Holt, Rinehart and Winston, 1963. Chapter 8.
3. Mendenhall, W., and L. Ott. *Understanding Statistics.* Belmont, Calif.: Duxbury Press, 1972. Chapter 5.
4. Slonim, M. J. *Sampling.* New York: Simon and Schuster, 1960.

Hypothesis testing

THE CALCULATED HUNCH

11

What is a hypothesis?

The term originally came from the term hypo-thesis or "sub-thesis," which meant it was derived from some more comprehensive or encompassing thesis, proposition, or theory. In this sense, testing an hypothesis has implications for a theory or general perspective. Put more generally, however, a **hypothesis** is a hunch. We all experience hunches every day. When stepping off the curb, you have a hunch about whether you will be hit. When taking a class, you have a hunch about whether you will pass. We can define a hunch as a guess about the actual outcome of something, given the data at hand. Likewise, a statistical hypothesis is a guess about the value of an outcome under certain conditions. The outcome is usually stated as a value of one or more parameters (i.e., measurement of a population). Thus, a statistical hypothesis is a statement about the value(s) in a population under certain conditions.

What is a null hypothesis?

The **null hypothesis** is the stated value of a parameter against which the empirical evidence is compared. It has become known as the null hypothesis because social scientists tend to state that any difference between a sample statistic and the population parameter is a sampling error and insignificantly different from zero. The word null comes from the Latin root meaning none, which is precisely what the hypothesis is assuming—no systemic difference. The symbol is the first letter of the word hypothesis with a zero as a subscript (H_0) indicating "not" or none.

What is the difference between an alternative and a null hypothesis?

An alternative or **research hypothesis** (as it is called by some social science researchers) is merely an alternative hunch about values of the parameters from which it may seem more plausible for the data to have come. Alternative hypotheses can be in the form of a specific alternative parameter as H_1: $\mu = 100$, or it may merely indicate that the mean X as an estimate of the population suggests μ is some unspecified alternative value such as H_1: $\mu \neq 100$. We usually compare the data from the sample with the statement in the null hypothesis, and come to some conclusion about whether it is more plausible to believe the alternative conditions are true or to tentatively accept the conditions stated in the null hypothesis. If our alternative hypothesis specifies direction, either by stating a value for the alternative or by using the greater or less than signs $>$, $<$ we are stating the alternative condition to be in one direction only. In such a case we utilize a **one-tailed test** of the null hypothesis. If the alternative hypothesis does not specify a direction for the alternative value, either by using the does not equal sign $\neq$ or stating alternative values to be on either side of the mean, we then use a **two-tailed test** of the null hypothesis.

How do we choose a critical region?

Ultimately any choice of cutoff is arbitrary. We call the size of the **critical region** alpha (α). This is the Greek a and corresponds to the percentage of all samples in the sampling distribution that would show values deviating from the hypothesized parameter, if the null hypothesis were true. It is specified as a probability (e.g., .05) and indicates a region in the tail of the sampling distribution. It is chosen in such a way as to make the probability of getting an extreme value of X very rare, and hence makes the choice of rejecting the null hypothesis a more confident one. If it is a two-tailed test, we divide the value of alpha in half and reject if the test statistic value falls in either end of the distribution (fig. 11-1).

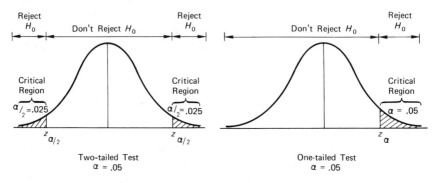

FIGURE 11-1

In the final analysis, the value chosen for alpha is going to depend on how willing we are to make a type I or type II error.

What is the difference between a type I and type II error?

A **type I error** *occurs if we reject the null hypothesis* (H_0) *when it is actually true.* The size of alpha (α) indicates the proportion of the time we risk making this type error. A **type II error** *is made by not rejecting the null hypothesis when it is false.* This is symbolized by beta (β). Alpha and beta are not directly proportional but do depend on one another. That is, the larger the value of α, the smaller the value of β, *given a specific value of the alternative hypothesis.* Since α is set before the test, it is up to the researcher to find the best test that will minimize beta, given that level of alpha (fig. 11-2).

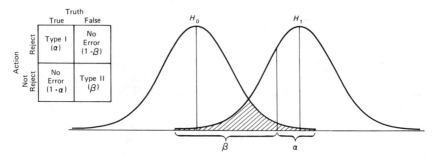

FIGURE 11-2

Think of the situation this way. You need a good used car and it has to last you a year. The dealer at Honest John's swears his car is in perfect shape even though you have doubts; another dealer has the same car which looks to be in better shape but costs twice as much. Which alternative would you pick? Obviously you have a dilemma. If you reject the cheaper car when it would have lasted the year, your decision ends up costing you twice the money; a type I error. On the other hand, if you buy the cheaper car and it does not last the year, you may end up spending more than twice as much in repairs; a type II error. This is a mundane example, but consider two cases that might be more critical.

You have discovered what you believe to be a vaccine against the common cold. If you set alpha at .05 in the animal screening tests, you are willing to take a 5% chance that you will conclude the vaccine is beneficial when in fact it is not. The rather large type I error in this context will assure that you will be less likely to reject the benefit of the vaccine (a type II error), but in the meantime you may be misleading the public about a cure. It would be important, therefore, not to commit a type I error, even if it means a greater type II error. Thus we would want alpha very low (e.g., .001).

On the other hand, here is an example of a situation in which it is important not to commit too large a type II error, even at the expense of a type I error. Suppose you are an instructor and somehow a copy of your final examination disappears before the day of the tests. You go ahead and administer the test anyway; now you must decide if it entered the hands of enough students to ruin the class distribution of grades. In this case, the null hypothesis is that there was no cheating; the alternative hypothesis is that there was. A type I error would be to assume that cheating took place (i.e., reject the null hypothesis) when in fact there was no cheating. A type II error would be to assume there was no cheating when in fact there was (i.e., falsely accepting the null hypothesis). Here you would want to minimize β. In fact, you would not mind doing so at the expense of a rather large type I error. Hence, you compute the class average and compare it to the standard in the population of students. Alpha is set very high (e.g., .25) and thus beta is low.

What criteria do we use to set alpha and beta?

First, it should be kept in mind that unless we know the specific value of the alternative parameter, we can never calculate beta. Unfortunately, most social scientists phrase hypotheses without knowing alternatives (e.g., $H_1: \mu \neq 75$). Thus, it becomes impossible to know a specific value of beta. Courts suffer somewhat the same dilemma. They may want to make very sure they do not send an innocent man to prison (type I error), and so they set up trials by peers and an elaborate appeal system to ensure against this. On the other hand, the court has no way of calculating the effect on society of letting a guilty man go free (type II error). Since the type I error is considered much worse according to our value system, in essence they try to obtain a low probability of that occurrence, even at the expense of increasing the likelihood of a type II error. Likewise, it is the nature of science that we do not want to flood the academic marketplace with false facts. Therefore, we are unwilling to take a very high risk of committing a type I error. The convention seems to be to set alpha very small (.05 and .01 have become tradition although the value will ultimately depend on the critical nature of the data) and perform the test by comparing the test statistic constructed from the sample with the test statistic associated with the value of alpha.

There may be a situation in which it becomes important to minimize both type I and type II errors (α and β). This can be done only by increasing the size of the sample. It should be clear that by doing this, we reduce the variance of error in estimating and hence the overlap of the distributions built around values in H_0 and H_1. Referring to the graph of alpha and beta, you should see that this will reduce both values.

What is the "power" of a statistical test?

A good test of our decision is indicated by how much of the time we reject the null hypothesis when it is false; that is, how often we are correct in rejecting. This is called the power of the test and is defined by the value $1 - \beta$. One minus the probability of *not* rejecting H_0 when it *is* false. In our previous used-car analogy, the strength of our judgment would lie in our ability to correctly spot defects in the cheaper car and reject its purchase.

The power of a test can take more than one value since we may pose our alternative hypothesis such that it can take on a variety of values (e.g., $H_1: \mu = 85$; $H_1: \mu = 87$, etc.). Since the power will be different for each alternative parameter, we often call it a power function. That is, it is a function of the value of the alternative. For a set value of alpha and N, all other things being equal, the power of a test will be greater the greater the difference between the value of H_0 and H_1. More will be said of this in chapter 15, on nonparametrics.

A good test of the power of our decision is indicated by how much of the time we reject the null hypothesis when it is false

Why do we say that we do *not reject rather than say we* accept *the null hypothesis?*

Whenever we test any hypothesis, we are testing it against stated alternative conditions which could seem more plausible. If the evidence is not strong enough to accept those stated alternative conditions, it does not necessarily mean that the null hypothesis is supported, only that the alternative is not. Another alternative hypothesis may be posed to account for the findings. If the alternative hypothesis proves to be weak and we accept the null hypothesis, we are guilty of what logicians call "affirming the consequent."

Think of this analogy. You go to your gym locker one afternoon excited to play tennis. Upon opening the locker door, you find your tennis racket is missing. After considering this situation, you come up with two hypotheses: H_0: your racket is at home versus H_1: someone has stolen your racket. Clearly if you could conclude that someone has stolen your racket (i.e., accept H_1), you would have to conclude it is not at home (i.e., reject H_0). However, just because you decide it has not been stolen (i.e., reject H_1) it does not mean that you must conclude that it is at home (i.e., accept H_0).

If you did, you would be guilty of affirming the consequent. Who is to say that you could not pose another alternative hypothesis which would better explain the findings. Perhaps you left the racket in the trunk of your car. Maybe you loaned it to a friend and forgot about it. Quite possibly the gym officials cleared out your locker. All these are plausible rival hypotheses. In social science, it is the purpose of research to eliminate alternatives one by one. Hence, we cannot accept the null hypothesis until *all* the possible alternatives have been tested.

What are the steps in testing hypothesis?

Not all tests are alike and some will involve different sorts of distributions, as we shall see; however, there are several processes involved in all hypothesis testing:

1. First we must state the hypothesis to be tested (i.e., the null hypothesis) together with alternative hypotheses which state a specific value or range of values. That is, we must determine the alternative decisions that can be made.

2. We must specify the appropriate sampling distribution to be used. If we are testing a sample proportion as an estimate of a population proportion, we must obtain the sampling distribution of the proportion, using the binomial. If we are testing a hypothesis about a population by using a sample mean, we must use the sampling distribution of the mean, which is a normal distribution.

3. We must select a critical region. This involves at least three things. First, we must decide whether the critical region is to be in one or two tails. If the alternative hypothesis expresses direction, it is a one-tailed test; if no direction is implied, it is a two-tailed test. Second, we must decide how much of a type I and type II error we are willing to make. This is going to be determined by how critical it is to reject the null hypothesis and be wrong vs. being more careful but not having a very powerful test. Third, as a corollary to this, we must determine the size of our sample. In general, the larger the sample, the smaller the probability of a type I and type II error. Sometimes cost and availability dictate the size of the sample; other times we can control the size and therefore must take into account when selecting a critical region.

4. We must perform the test. This involves computing the actual test statistic (such as a z-score) and comparing it with the cutoff value (such as the z-score) corresponding to the alpha value. On the basis of this comparison a decision is made to either reject or not reject the null hypothesis. In general, if the test statistic exceeds the value associated with the cutoff point, we reject the null hypothesis. If it

does not exceed the value associated with the cutoff point, we do not reject.

> *Is the hypothesis-testing procedure followed in every
> inference from sample statistics?*

Yes. We will see as we progress into further topics, the type of sampling distribution may change and the form of the hypothesis may be slightly different, but the procedure is the same. In the next chapter we begin by discussing tests of hypotheses using the normal sampling distribution when making inferences about a population mean on the basis of data from a single sample.

How Much Do You Remember?

New Words	*New Symbols*
Hypotheses	H_0
Null	H_1
Research	α
One and two-tailed tests	β
Critical region	
Type I and II errors	
Alpha	
Beta	
Test statistic	
Power	

Did You Ever Wonder?

1. Why is the alternative hypothesis called the research hypothesis?
2. What conditions might make a type II error worse than a type I error?
3. Why can we never "accept" the null hypothesis when testing a nonspecific alternative hypothesis?

Want to Know More?

1. Hays, W. L. *Statistics.* New York: Holt, Rinehart, and Winston, 1963. Chapter 9.
2. McCall, R. B. *Fundamental Statistics for Psychology.* 2nd ed., New York: Harcourt Brace Jovanovich, Inc., 1975. Chapters 8–9.

3. MENDENHALL, W., and L. OTT. *Understanding Statistics*. North Scituate, Mass.: Duxbury Press, 1972, Chapter 6.

4. MORRISON, D., and R. HINKEL. *The Significance Test Controversy*. Chicago: Aldine Publishing Company, 1970.

The single sample test

PRECISION IN THE FACE OF AMBIGUITY

12

*How do we know anything about populations,
especially when they are very large?*

Many indices exist in social science to help us determine population values. We have access to a census that is taken every ten years. We also have access to studies over time which show trends. Indeed, tools such as time-series analysis help establish this kind of knowledge. Often, we make use of these censuses and trends to establish population parameters, that is, means and variances of a particular trait in the population.

*If the mean of a population is already known,
what is the purpose of using the sample to test
a hypothesis about that population mean?*

There are two reasons, really. In the first case, we want to test to see if our sample is representative of the population from which it was drawn (e.g., does $\bar{X}$ represent μ). Think of the times when you have had to decide which class to take in making up your class schedule. If you had some information from a student evaluation, for instance, about the grades given to the students in your department over the years and you were able to look at the grades in a particular class as well, you might be interested in knowing whether the grades in that class were representative of grades given by the department as a whole. Or put more simply, if you were looking for an easy class to include in your schedule and you knew the department had the reputation of assigning easy grades, you might want to "test" whether a certain professor's grading is as easy as that of the rest of the department; that is, do his grades represent a random selection of grades from the entire department? This could be done by comparing the mean grade point average in the professor's class (the single sample) to the mean of all grades given in the department (the population). Therefore, using information about the sample of this professor's grades, we can infer whether he is representative of the grading in the entire department.

A second use of this test might be to show that within a given degree of error (α), we can assume this sample is from a different population. For instance, if you are a female, you no doubt wear pantyhose. Long years of experience have shown that under average conditions the hose tend to last a certain length of time before the first run appears. Now suppose a new brand comes on the scene that guarantees an unbelievable resistance to runs. So you buy a sample

You might want to test that a certain professor's grading
is as easy as that of the rest of the department

of several to test the manufacturer's claim. The null hypothesis is that these new hose will be no different than the old. The alternative is that their claim is true. What you really would like to conclude is that the new brand is indeed superior, thus ending the frustration over getting runs. We can make a test by comparing the mean length of time it takes to get runs in our sample of new pantyhose to the mean length of time established over the years.

These two examples of using the test by comparing a sample mean with the population mean may sound vaguely alike, as indeed, they are. In both circumstances, we know the population mean and the population variance. The only difference is that in the first case, we are trying to justify the fact that our sample selection (grades by the easy professor) is representative of the population (grades in an easy department). In the second case we are interested in showing, not that our sample is representative, but rather that it is significantly different from the population.

What role does the sampling distribution of the mean play in testing a hypothesis about the population means?

Remember that the expected value of the sampling distribution of the mean is $\mu_{\bar{X}}$; that is, if we theoretically took an infinite number of repeated samples and plotted their means, the resulting normal distribution would have a mean equal to the population parameter $\mu_{\bar{X}}$. Therefore, if we have information about the population, we are in a position to compare any given sample to it and specify the amount of error we are making in using our sample to estimate the population mean μ. This is done by determining the position of our sample mean in the sampling distribution of the mean. In effect, we are determining a z-score for our sample mean. It shows the relative position and therefore the proportion of the time that we would attain a sample mean deviant to that degree.

Remember, a z-score is defined as a value minus its mean, divided by its standard deviation. In this case, the value is the mean of the sample. Its mean is the mean of the sampling distribution, which is the same as the population mean μ. The standard deviation of the distribution is the standard error of the mean $\sigma / \sqrt{N}$. Thus, the z-score, or relative position of the sample mean, is

$$z = \frac{\bar{X} - \mu}{\sigma_{\bar{X}}} = \frac{\bar{X} - \mu}{\sigma / \sqrt{N}}$$

As we stated earlier, this represents a position in the standard normal distribution. We can think of the z-score percentage associated with this position as the probability of randomly selecting for a sample mean with that value.

What is the hypothesis testing procedure when single samples are compared to populations?

Suppose you plan to start a retail outlet for bicycles in the west area of town. Checking the census establishes the mean age of the area to be 38 years with a standard deviation of 8 years. Since bicycles appeal to the young, you begin to doubt that this would be a good place to start the store. However, you have reason to believe that the age composition has been changing recently. If you draw a sample of people from the area and ask their ages, a single sample test could be performed to see whether any significant changes have taken place since the census.

The first step is to state the hypotheses. In this case the null hypothesis is H_0: $\mu = 38$. The alternative is H_1: $\mu < 38$. The next step is to specify the sampling distribution and statistic. Since we are assuming that age is normally distributed and are testing a mean, we can use the z test and utilize the sampling distribution of the mean. Third, we must select a critical region. Since the decision to build the store involves a large sum of money, we may want to be very sure in the decision to reject H_0. Thus, we might choose alpha = .01. Because alternative hypothesis is directional, the test is one-tailed. This means that the size of the critical region α is in one tail only. Our sample size is set at 36. The last step is to construct the z value and compare it with the z value associated with alpha = .01:

$$\bar{X} = 35 \quad \mu = 38 \quad \sigma = 8 \quad N = 36$$

$$z = \frac{35 - 38}{8/\sqrt{36}} = \frac{-3}{8/6} = \frac{-18}{8} = -2.25$$

$$z_{.01} = 2.33 > |2.25|$$

Since the absolute value of our constructed z score is less than the z value associated with an alpha of .01, we do not reject the null hypothesis, and must tentatively conclude that the mean age of the neighborhood is the same as it was in the past.

*Do we always know the population variance? If not,
how can a z value be calculated?*

Sometimes we do not know the population variance. For
instance, if studies have indicated a stable rate over time, we may not
have access to the raw data and therefore cannot compute the variance
in the population. In this case, we have to make use of the best available
evidence, that is, information from the sample. We compute the standard
error of the mean by replacing the variance of the population with the
sample estimate:

$$\text{Estimated } \sigma_{\bar{x}} = \frac{\hat{s}}{\sqrt{N}} \quad \text{where} \quad \hat{s} = \sqrt{\frac{\Sigma(X_i - \bar{X})^2}{N-1}} \, .$$

The symbol ^ stands for an unbiased estimate of the population standard
deviation. It sounds tempting to merely replace σ by s in the formula
for the standard error. However, it so happens that this is a consistent
underestimate of the population variance and therefore constitutes bias.
To compensate for this, we multiply the sample variance s^2 by $N/N - 1$.
The result is an unbiased estimate of σ^2:

$$\text{Thus } \hat{s}^2 = s^2 \cdot \frac{N}{N-1} = \frac{\Sigma(X_i - \bar{X})^2}{\cancel{N}} \cdot \frac{\cancel{N}}{N-1} = \frac{\Sigma(X_i - \bar{X})^2}{N-1}$$

Because this is true, we can restate the estimated standard error as

$$\text{Estimated } \sigma_{\bar{x}} = \sqrt{\frac{\hat{s}^2}{N}} = \sqrt{\frac{\Sigma(X_i - \bar{X})^2}{N-1} \cdot \frac{1}{N}}$$

$$= \sqrt{\frac{\Sigma(X_i - \bar{X})^2}{N} \cdot \frac{1}{N-1}} = \sqrt{\frac{s^2}{N-1}} = \frac{s}{\sqrt{N-1}} = s_{\bar{x}}$$

This looks vaguely like the standard error of the mean with σ replaced
by s. However, the important difference is that we are now dividing
by $N - 1$ rather than N. This $N - 1$ corrects for the underestimate
of the population variance. As you can see, when N is large (e.g.,
over 30), the correction is almost negligible and we may resort to the
z ratio as our test statistic. However, when N is small the correction
can have considerable effect. In these cases, we must use the **t ratio**
as the test statistic and rely on the t distribution for the table of
probabilities.

How does the t *distribution differ from a* z *distribution?*

William Gosset's statistics for small samples are known as
t distributions, or student's *t,* after his pen name of "student" (see
appendix I). The distributions rest on a fundamental observation that
as *N* gets smaller, the effect of the unbiased estimate of the population
variance becomes more important. Thus, when *N* is large, the *t* distribution is virtually the same as the *z* distribution. When *N* is small, however,
the *t* distribution tends to flatten out at the ends, indicating the larger
estimated standard error. $N - 1$ in our estimate is the **degrees of freedom**.
(fig. 12-1). Degrees of freedom is a rather simple concept, and relates
in this case to the computation of the standard deviation. Since in
computing this value we subtract the mean from the individual scores,
and the mean is known, specifying all but one of the deviations would

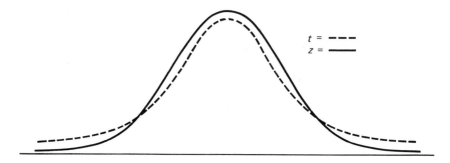

FIGURE 12-1

allow us to determine the last one. In a sense, we are saying that $N - 1$
of these values are free to vary before the last one is determined. In
general, the degrees of freedom is equal to the number of values *N*
minus the number of parameters being tested. In this case, we are testing
one population mean; hence, the degrees of freedom (abbreviated d.f.)
is $N - 1$. Therefore, the *t* ratio is defined in much the same way as
the *z* value except that the standard error has been estimated:

$$t = \frac{\bar{x} - \mu}{s/\sqrt{N - 1}}$$

We are able to look up alternative *t* distribution (corresponding to different
d.f.) in one table since only selected values of α have been included
(see appendix IV.). We select the value of alpha across the top of

the table depending on whether it is a one- or two-tailed test. The degrees of freedom are found down the side. The value listed under this column and row is the t value associated with the probability that we would achieve a sample mean this extreme. As with the z ratio, if our constructed t value exceeds that found in the tables, we reject the null hypothesis. If not, we do not reject.

Take your grade point average, for example. Suppose this term you randomly selected five friends and computed their grade point, finding a 2.9 average. A published report listed the overall school average as 2.2 for the term. Could you test the null hypothesis (H_0) that your group was no different from the average versus the alternative (H_1) that they did perform better, if you did not know the population standard deviation but did know that your sample s was .55?

$$\mu = 2.2 \qquad N = 5$$
$$\bar{X} = 2.9 \qquad \alpha = .05$$
$$s = .55 \qquad \text{d.f.} = 4$$

$$t = \frac{2.9 - 2.5}{\dfrac{.55}{\sqrt{4}}} = \frac{.7}{\dfrac{.55}{2}} = \frac{1.4}{.55} = 2.54$$

$$t_{.05;4} = 2.132 \qquad 2.54 > 2.132$$

Since the computed t value is greater than the critical t value, you reject the hypothesis of no difference and accept the alternative that your friends performed better.

What conditions determine use of the z distribution versus the t distribution?

It sounds as if we use the t distribution whenever we do not know the population variance. However, we have stated that it is virtually equal to the z distribution when the N is large. As a rule of thumb, the t ratio is a "small sample" statistic. To be reasonably accurate, you need use the t ratio only if the population variance is unknown *and* the size of the sample is less than 30. However, if the t statistic is utilized, one additional restriction need be considered. We must assume that the population distribution is normally distributed. In practice, however, this assumption is often violated, with little resulting error. More on this will be discussed in chapter 15, on nonparametrics.

Does the single-sample procedure apply to the binomial distribution?

Yes. We know that the expected value of the binomial distribution is NP where N is the number of trials and P is the proportion of occurrence. Further, we know that when N is large, the binomial distribution is approximately normal (see chap. 10). It should seem reasonable, therefore, that we can construct a z value for the binomial case. The value is the number of favorable outcomes in our sample X; the mean is the expected value of the distribution NP; and the standard deviation is $\sqrt{NPQ}$. Thus, the z value is

$$Z_p = \frac{X - NP}{\sqrt{NPQ}}$$

It may be a bit confusing to see X as the *number* of favorable outcomes when we are speaking of proportions. With a small amount of rearrangement in the formula (i.e., we divide everything in the formula by N) we can see that this is really a single-sample test of proportions:

$$z_p = \frac{X - NP}{\sqrt{NPQ}} = \frac{(X/N) - (NP/N)}{\dfrac{\sqrt{NPQ}}{N}} = \frac{P_s - P_\mu}{\dfrac{\sqrt{N}}{\sqrt{N}} \cdot \dfrac{\sqrt{PQ}}{\sqrt{N}}}$$

$$= \frac{P_s - P_\mu}{\sqrt{\dfrac{PQ}{N}}}$$

where P_s is the sample proportion and P_μ the population proportion.

Take the example of the National Basketball Association Championships when an announcer for the Los Angeles Lakers was complaining that the free-throw percentage of one of the star players was much worse during the playoffs than his career total indicates. We can test the validity of this accusation by checking the record book. The playoff percentage is the sample proportion and the career percentage is the population proportion. Suppose the records showed that our star made 320 out of 400 shots during his career but only 5 of 9 during the playoffs. We would have the following information:

$$P_\mu = {}^{320}/_{400} = .80 \qquad P_s = {}^5/_9 = .55$$

$$Q_\mu = 1 - P_\mu = .20 \qquad N = 9$$

We construct our z value in the manner outlined above and compare it with the value associated with alpha. Suppose we choose alpha =

.01, we make the decision as follows:

$$z = \frac{P_s - P_\mu}{\sqrt{P_\mu Q_\mu / N}} = \frac{.55 - .80}{\sqrt{(.8 \times .2)/9}} = \frac{-.25}{\sqrt{.16/9}} = \frac{-.25}{.4/33}$$

$$= \frac{-.75}{.4} = -1.89$$

$$Z_{.01} = 2.33 \qquad |1.89| < 2.33$$

Since our value is less than that found in the z table, we do not reject the null hypothesis and must assume that the announcer was appealing to emotion rather than to statistical fact!

Can we test a hypothesis about a population mean using a single sample if the value of the population mean is unknown?

We cannot test a hypothesis under these circumstances since we cannot specify any value for the population mean being tested. Instead, we become interested in knowing how confident we can be in estimating an interval in which we can be sure that the value of the population mean can be found. Logically, we use the term **confidence interval.** This can be found by merely solving the z ratio for μ:

$$\pm z = \frac{\bar{X} - \mu}{s_{\bar{X}}}: \qquad \pm z \cdot s_{\bar{X}} = \bar{X} - \mu$$

Therefore $\mu = \bar{X} \pm z s_{\bar{X}}$

The z value may be used to find the probability that an interval so

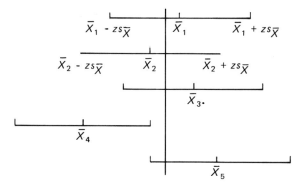

FIGURE 12-2

constructed includes the population mean; $\bar{X} + zs_{\bar{x}}$ and $\bar{X} - zs_{\bar{x}}$ are the upper and lower limits of that interval. In other words, if we took an infinite number of samples from the population and constructed an interval around each sample mean, our sampling distribution of the means would tell us that the value of the population mean would be contained in a certain percentage of the sample intervals. In general, that percentage of samples corresponds to $1 - \alpha$ for a two-tailed test. (fig. 12-2). For instance, you have the following sample data and want to know the limits of the interval in which you can be 95 percent sure the population mean is contained (figure 12-3). Obviously, the interval for a t value

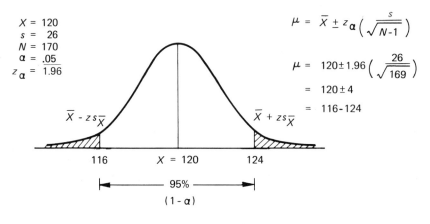

$X = 120$
$s = 26$
$N = 170$
$\alpha = .05$
$z_\alpha = 1.96$

$$\mu = \bar{X} \pm z_\alpha \left(\frac{s}{\sqrt{N-1}} \right)$$

$$\mu = 120 \pm 1.96 \left(\frac{26}{\sqrt{169}} \right)$$

$$= 120 \pm 4$$

$$= 116 - 124$$

$\bar{X} - zs_{\bar{X}}$ $\bar{X} + zs_{\bar{X}}$

116          $X = 120$ 124

95%
$(1-\alpha)$

FIGURE 12-3

is constructed in the same way except that we must look up the t value in accordance with the degrees of freedom.

You might be curious about what is done when we are ignorant of the mean or standard deviation of the population and we aren't interested in estimating a confidence interval? What tools do we have to explain the difference that we have observed between one sample and other? This is our next topic.

How Much Do You Remember?

New Words	New Symbols
z ratio	z
Degrees of freedom	d.f.
t ratio	$s_{\bar{x}}$ or $\hat{s}$
Confidence interval	t

Did You Ever Wonder?

1. In terms of probability is there a difference between estimating a population mean and establishing a confidence interval?
2. Why must we assume the population distribution is normal when using the *t* test?
3. Why should the standard score for the binomial be approximately normal?

Want to Know More?

1. JOHNSON, R. R., *Elementary Statistics.* North Scituate, Mass.: Duxbury Press, 1973. Chapters 8–9.
2. HABER, A., RUNYON R., and BODIA, P. *Readings in Statistics.* Reading, Mass.: Addison-Wesley Publishing Company Inc., 1970. Esp. Block 2.
3. MUELLER, J. H., K. F. SCHUESSLER, and H. L. COSTNER. *Statistical Reasoning in Sociology.* 2nd Ed. Boston: Houghton-Mifflin Company, 1970.
4. RUYON, R. P. and A. HABER. *Fundamentals of Behavioral Statistics.* 2nd ed. Reading, Mass.: Addison-Wesley Publishing Company, 1971. Chapter 12.

Two samples

HOW DIFFERENT
IS DIFFERENT?

13

How can we test the difference between two population means?

Think of this analogy: Suppose you are at a wedding reception one hot muggy Sunday afternoon at which large groups of people are congregating around two punch bowls spiked with vodka. You would like to be able to take advantage of the refreshment from both bowls, as the lines are long, but you are afraid each may contain greatly differing amounts of alcohol. Mixing different concentrations of liquor has always made you very sick, so rather than bother the host or chance unpleasant aftereffects you decide to run your own test.

You ask several people surrounding each bowl to taste their punch and tell you how much vodka they believe is in the drink. On the basis of information about the average concentration of alcohol in these two samples, you must make a decision whether the amount of liquor in both bowls is actually the same or whether there are two significantly different concentrations being served.

Now translate this analogy into numbers. Consider two independent samples (typically nonquantitative categories such as black-white, male-female, or Catholic-Protestant) within which have been measured some interval-level trait (such as I.Q. scores). Testing the difference of means in those two samples can establish whether or not that difference is

We are testing the difference between two sample means
to infer whether or not the two populations from which
they are drawn have the same mean

large enough for us to conclude that the two groups come from different populations.

What we are doing in this case is testing the difference between two sample means to infer whether or not the two populations from which they are drawn have the same mean. The null hypothesis (always stated in terms of population parameters) is that there is no difference between population means $H_0: \mu_1 = \mu_2$ or $\mu_1 - \mu_2 = 0$). We could hypothesize that the difference is any value; most often, however, we are interested in testing to see whether the two populations have identical means.

Is there a sampling distribution of means for testing the difference of two population means?

Yes! You should remember that the central limit theorem tells us that any combination of independent variables repeatedly and randomly sampled tends to be normally distributed the larger the sample size N. At first, we used this information to assert that combining terms from a sample, by computing the mean, would yield a sampling distribution of means which is normal. Now we are in a position to extend that theorem. If we continue to draw pairs of samples, and compare the means from each pair by subtracting one from another, these differences will still be a legitimate combination of independent random variables. Because this is true, the distribution of the differences derived by comparing all possible pairs of sample means will approximate a normal distribution. Such a distribution is called a **sampling distribution of the difference of means.** It is normally distributed in terms of area under the curve. Thus, we can test an hypothesis about the populations using the sample data, much the same way that we did for a single sample. If the null hypothesis states that the two population means are identical, $H_0: \mu_1 = \mu_2$, we are saying essentially that the two samples come from the same population. The alternative hypothesis is that they are different populations; that is $H_1: \mu_1 \neq \mu_2$ or $\mu_1 < \mu_2$ (fig. 13-1).

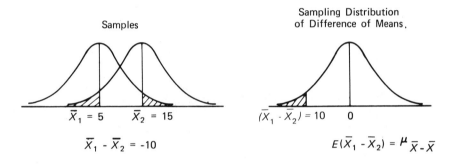

FIGURE 13-1

Under the null hypothesis the expected value of the sampling distribution of the difference of means is zero (i.e., the differences average out to zero). Therefore, $\mu_{\bar{X}_1 - \bar{X}_2} = 0$. The standard deviation of these differences is

$$\sigma_{\bar{X}_1 - \bar{X}_2} = \sqrt{\frac{\sigma_{\bar{X}_1}^2}{N_1} + \frac{\sigma_{\bar{X}_2}^2}{N_2}},$$

a new set of symbols, and is called the **standard error of the difference** (of means).

How do we construct a test statistic for the difference of population means?

Our sampling distribution of differences is normally distributed. So as long as we know the value, the mean of the values, and the standard deviation of our value, we can construct a z score which we then compare to the z score associated with the critical region α.

The value in this case is the difference in our two sample means. Under the null hypothesis, the mean of those values is the mean of the sampling distribution of differences $\mu_{\bar{X}_1 - \bar{X}_2}$, which is zero. The standard deviation of the distribution is the standard error of the difference of means. This is symbolized by $\sigma_{\bar{X}_1 - \bar{X}_2}$. Hence, the numerator of our z score is $(\bar{X}_1 - \bar{X}_2) - (\mu_1 - \mu_2)$. But $\mu_1 - \mu_2$ is zero and so the term drops out. Note that

$$\sigma_{\bar{X} - \bar{X}} = \sqrt{\frac{\sigma_{X_1}^2}{N_1} + \frac{\sigma_{X_2}^2}{N_2}}$$

and so our test statistic is a z value, which is

$$z = \frac{\bar{X}_1 - \bar{X}_2}{\sqrt{\dfrac{\sigma_{X_1}^2}{N_1} + \dfrac{\sigma_{X_2}^2}{N_2}}}$$

We compare this value to the criterion z score in exactly the same way we did for the single-sample and make our decision to reject H_0 if our z score exceeds that associated with the chosen critical region (i.e., alpha).

Why do we add *the two standard errors of the mean to derive the standard error of the* difference *of means?*

Remember that the standard error $\sigma_{\bar{X}}$ is just the square root of the variance of that error $(\sigma_{\bar{X}}^2)$. Whenever we combine squared terms, we must weight them by the square of their coefficients. Since we are taking one unit of variance from the first population and subtracting from it one unit of variance from the second population, the respective weights are $(1)^2 + (-1)^2$. Hence we have:

$$(1)^2 \frac{\sigma_1^2}{N_1} + (-1)^2 \frac{\sigma_2^2}{N_2} = \frac{\sigma_1^2}{N_1} + \frac{\sigma_2^2}{N_2}$$

The standard error of the difference is just the square root of this:

$$\sqrt{\frac{\sigma_1^2}{N_1} + \frac{\sigma_2^2}{N_2}}$$

What if the population variances are not known?

When we know the population variance or when the N is large (about 30), we can use z values. Unfortunately neither of these requirements are met very often. Indeed, if we know the variance of the two population distributions we would already know their means (since we must use them to compute the variance). If we knew the population means, it would be absurd to hypothesize about their difference. We would know the exact difference by merely subtracting them!

Even if we have to estimate the variances from the sample data, however, we know that the z score will still be usable if the N in the samples is large enough. Unfortunately, in comparing groups, this may also occur infrequently. In fact, social scientists often deal with very small samples in making decisions about differences.

These two facts coupled together mean that we will often be dealing with the t distribution for the difference-of-means tests. This should not bother us, since it provides for a better decision when the N is small. To use the t distribution we merely plug in our estimate of the variances to achieve the t value:

$$t = \frac{(\bar{X}_1 - \bar{X}_2) - (\mu_1 - \mu_2)}{s_{\bar{X}_1 - \bar{X}_2}} = \frac{\bar{X}_1 - \bar{X}_2}{\sqrt{\dfrac{\hat{s}_1^2}{N_1} + \dfrac{\hat{s}_2^2}{N_2}}}$$

$$= \frac{\bar{X}_1 - \bar{X}_2}{\sqrt{\dfrac{s_1^2}{N_1 - 1} + \dfrac{s_2^2}{N_2 - 1}}}$$

Remember that the t distribution is actually a family of distributions and that we must look up the value based on the number of degrees of freedom. In the two-sample case, we have $N - 1$ values that can vary in each sample. Hence, the degrees of freedom is $(N_1 - 1) + (N_2 - 1) = N_1 + N_2 - 2$.

How do we compute the estimated standard error of the difference?

If we can assume along with our null hypothesis ($\mu_1 = \mu_2$) that the variances of the two populations are equal ($\sigma_1^2 = \sigma_2^2$); that is, that there is one and only one population distribution with mean $\mu_1 = \mu_2$

$= \mu$ and variance $\sigma_1^2 = \sigma_2^2 = \sigma^2$ from which our two samples have been drawn, we merely pool the two sample variances (i.e. take a weighted average of the variances) to estimate the standard error:

$$t = \frac{\bar{X}_1 - \bar{X}_2}{\sqrt{\dfrac{\hat{s}_1^2}{N_1} + \dfrac{\hat{s}_2^2}{N_2}}} = \frac{\bar{X}_1 - \bar{X}_2}{\sqrt{\left(\dfrac{1}{N_1} + \dfrac{1}{N_2}\right) \cdot \hat{s}^2 \text{ pooled}}}$$

$$\text{but} \quad \hat{s}^2 = \frac{s^2}{N-1}$$

$$\text{therefore} \quad t = \frac{\bar{X}_1 - \bar{X}_2}{\sqrt{\left(\dfrac{1}{N_1} + \dfrac{1}{N_2}\right)\left(\dfrac{N_1 s_1^2 + N_2 s_2^2}{N_1 + N_2 - 2}\right)}}$$

For example, assume you want to test the notion that girls in your school go out more often than boys. You ask a random sample of females and males (from your school) how many times a month they go out, receiving these data:

	Females	Males
	$\bar{X}_1 = 7$	$\bar{X}_2 = 5$
	$s_1 = 3$	$s_2 = 2$
	$N_1 = 17$	$N_2 = 26$

$$H_0: \mu_{X_1} = \mu_{X_2} \quad H_1: \mu_{X_1} > \mu_{X_2}$$

$$t = \frac{7 - 5}{\sqrt{\dfrac{1}{17} + \dfrac{1}{26}\left[\dfrac{17(9) + 26(4)}{41}\right]}} = \frac{2}{\sqrt{\left(\dfrac{43}{442}\right)\left(\dfrac{257}{41}\right)}}$$

$$= \frac{2}{\sqrt{(.097)(6.3)}} = \frac{2}{\sqrt{.61}} = \frac{2}{.78} = 2.56$$

$$\text{d.f.} = N_1 + N_2 - 2 = 17 + 26 = 41 \quad t_{.05,41} = 1.684$$

$2.56 > 1.684$ therefore reject H_0.

We reject the hypothesis that there is no difference in dating and accept the alternative that girls do have more dates.

What if we do not know the population variances are equal?

Suppose a sampling distribution is composed of variances (i.e., average squared distances from the mean) rather than individual values. We can test the error of any one sample's variance in estimating the population variance, much the same as we do with a sample mean in estimating the population mean. However, since we are dealing with squared values, it should stand to reason that the sampling distribution will not be *linear* combination of terms, and therefore will have a slightly different than normal shape. Such a distribution of squared values is called a chi square distribution and is denoted by the symbol χ^2. More specifically, **chi square** *is a distribution of the ratio of the sample sum of squares to the actual population variance.* Hence

$$\chi^2 = \frac{(N-1)\,\hat{s}^2}{\sigma^2} = \frac{\Sigma\,(X - \bar{X})^2}{\sigma^2}$$

In order to test the hypothesis that two populations have equal variances (i.e., H_0: $\sigma_1^2 = \sigma_2^2$) we would want to compare each sample's estimate of its respective population variance. This is done by looking at the difference (i.e., computing a ratio) of the two chi square distributions, each divided by its degree of freedom. If H_0: is true, that the two populations have equal variances, this should reduce to

$$\frac{\dfrac{\chi_1^2}{\text{d.f.}}}{\dfrac{\chi_2^2}{\text{d.f.}}} = \frac{\dfrac{\hat{s}_1^2}{\sigma_1^2}}{\dfrac{\hat{s}_2^2}{\sigma_2^2}} = \frac{\hat{s}_1^2}{\hat{s}_2^2}$$

This ratio of two sample variances under the null hypothesis H_0: $\sigma_1^2 = \sigma_2^2$ is distributed according to an *F* statistic (after its creator R. A. Fisher). Actually, the **F distribution** *is defined as the ratio of two chi square distributions each divided by their respective degrees of freedom*, and is used to test the ratio of any two variances to see if one is significantly different from the other. That is,

$$\text{Reject } H_0: \sigma_1^2 = \sigma_2^2 \quad \text{if} \quad \frac{\hat{s}_1^2}{\hat{s}_2^2} \geq F \text{ value for a given alpha } (\alpha)$$

The *F* distribution also has a direct relationship to the *t* distribution. In fact the square of the *t* ratio is only a special case of the *F* ratio. You should satisfy yourself that the value found in the *F* table is identical with the t^2 value when the degrees of freedom for the numerator equals 1 (i.e., one difference).

It is not the purpose of this book to derive the general case of F as a test of the ratio of variances from the specific case of the sampling distribution of variances, although it can be shown (see references at the end of the chapter). Rather, by getting an intuitive feel for the relation between F, χ^2, and t, we can see that to use the F distribution to test our assumption about the equality of variances in a difference of means test, we merely calculate the ratio of the two sample estimates of the population variances (*always putting the larger one on top to assure a value greater than 1*). We then look up the value in the probability distribution of F to see if they are different enough from each other to consider their respective population variances unequal. If they are not, we assume that the population variances are the same.

As with the t table, we must look at our F table by specifying the number of degrees of freedom. Since both variances in the F ratio are estimating a value, we have two different degrees of freedom, corresponding to the two sample variances ($N_1 - 1$ for the first sample; $N_2 - 1$ for the second sample). Looking up the F value, we see that its probabilities are listed for only the .05 and .01 levels. Obviously, we could list the values for any significance level we desire as with t or z, but this would take an entire book. (Incidentally, it is not by historical accident that social scientists, quoting the Fisher and Yates' original statistical tables for alpha values of .05 and .01, have contributed significantly to their establishment as traditional values of alpha.) At any rate, for a given level of probability (e.g., .05) we look up the critical F value (see Appendix IV) by reading the degrees of freedom of the larger variance across the top, and the degrees of freedom of the smaller variance down the side. The value corresponding to the intersection of that row and column will be the F value associated with the level of significance α listed at the top of the table. Comparing the computed F statistic with the value looked up, the decision is the same: if our value does not exceed the value in the table, we assume that the population variances are the same. We accept that they are different if our value does exceed that found in the table.

How do we combine estimates of the population variances if we cannot assume that they are equal?

If we cannot assume equality of variance (called **homoscedasticity**), we are essentially testing the null hypothesis that the population means are equal, though coming from distributions with different variances.

In this case, we cannot simply pool our estimates of the variance for the sampling distribution of the difference of means. We must estimate them separately. Thus, the t ratio is

$$t = \frac{\bar{X}_1 - \bar{X}_2}{\sqrt{\dfrac{s_1^2}{N_1 - 1} + \dfrac{s_2^2}{N_2 - 1}}}$$

This is a more general statement of our t statistic and therefore is more conservative. Given the same data, this formula will typically yield a t value slightly less than the t value using the pooled estimate. However, we must also use a more conservative estimate of the degrees of freedom. In our previous campus dating example, this value drops from 41 to approximately 25. The t value associated with alpha = .05 and 25 degrees of freedom is now 1.708, a slightly more demanding cutoff than 1.684. Therefore, if we recompute our example (p. 137) we find,

Females	Males
$\bar{X}_1 = 7$	$\bar{X}_2 = 5$
$s_1 = 3$	$s_2 = 2$
$N_1 = 17$	$N_2 = 26$

$$t = \frac{7 - 5}{\sqrt{9/16 + 4/25}} = \frac{2}{\sqrt{.56 + .16}} = \frac{2}{\sqrt{.72}}$$

$$= \frac{2}{.85} = 2.35 \qquad t_{.05;25} = 1.708$$

$$2.35 > 1.708 \quad \text{therefore reject } H_0$$

In each case, we are stating the null hypothesis of no difference between population means (H_0: $\mu_1 = \mu_2$). The advantage of being able to assume equal population variances and use a pooled estimate of the variance is that we produce a more powerful statistic and do not require the restriction on the degrees of freedom.

Is there any reason to bypass using an F test to test the equality of population variances?

In general, we can violate the assumption of equal population variances without too much error if the N's of each sample are rather large. Some statisticians even argue that the test for homogeneity of variance is senseless given the small amount of error that is apt to result by assuming equal variance.

Even if N's are not large, another way to minimize the difference between these two estimates is to draw samples with equal N's from each population. If this is done, we can demonstrate that the formulae are essentially the same; the only difference is our estimate of the degrees of freedom:

$$\sqrt{\left(\frac{1}{N_1} + \frac{1}{N_2}\right)\left(\frac{N_1 s_1^2 + N_2 s_2^2}{N_1 + N_2 - 2}\right)} = \sqrt{\frac{s_1^2}{N_1 - 1} + \frac{s_2^2}{N_2 - 1}} \quad \text{if } N_1 = N_2$$

Since $N_1 = N_2$,
$$\left(\frac{1}{N_1} + \frac{1}{N_2}\right) = \frac{2}{N}$$

and
$$\left(\frac{N_1 s_1^2 + N_2 s_2^2}{N_1 + N_2 - 2}\right) = \frac{N(s_1^2 + s_2^2)}{2(N - 1)}$$

Therefore $s_{\bar{X} - \bar{X} \text{pooled}} = \sqrt{\left(\frac{2}{N}\right)\left(\frac{N(s_1^2 + s_2^2)}{2(N - 1)}\right)}$

$$= \sqrt{\frac{s_1^2 + s_2^2}{N - 1}} = \sqrt{\frac{s_1^2}{N_1 - 1} + \frac{s_2^2}{N_2 - 1}}$$

By securing equal N's, we help suppress the problem of assuming equal population variances.

Which formula for t is most likely to be used?

They are used in analyses of both survey and experimental data. In surveys, however, we typically collect information in a naturally occurring setting and are therefore bound by the number of responses we receive. Throwing cases out, even at random, will affect the number of degrees of freedom. Thus, it is safer to use the more conservative estimate of the standard error to assess differences, since the size and variance of the samples are liable to be unequal.

Experimental settings, on the other hand, are more likely to yield equal N's and variances since the experimenter decides how many subjects he will use and where they will be placed. Generally, we have two groups, one of which receives a treatment (**the experimental group**) and one of which does not (**control group**). Because it is assumed that our subjects are drawn and assigned to groups at random, the only difference between the groups is the treatment. Thus, if we find a significant difference between our sample means, we are in a position to attribute that difference to the treatment, and assume that this effect would hold for the population in general. Because of the likely homogeneity of variances, the pooled estimate of the standard error is often used for *t* tests in experimental situations.

Are there experiments where we would not use random assignment of subjects in testing the difference of means?

Yes. Rather than testing two groups, I might want to test the difference of scores obtained in one group on two occasions. A good analogy might be checking to see if the members of your Girl Scout troop sold the same number of boxes of cookies this year as they did the year before, or whether each class in a grammar school collected the same quantity of newspapers in the paper drive this year as last. In these cases, the two samples are not independent since the same persons are measured on both occasions. The classic example is the before and after experimental design, the treatment being given between measurements.

What we are testing in that case is the difference of scores from the first occasion to the next. Although the scores are correlated, the differences are independent random variables. Since the distribution of the differences in scores from one occasion to the next is approximately normal, we can compute a test statistic based on *t*. The value is the mean difference (let *D* stand for difference) across individuals in the sample; the mean of the distribution is the expected difference, which according to the null hypothesis is zero; and the standard error is the standard deviation of the difference:

$$t = \frac{\bar{D} - \bar{X}_D}{\sqrt{s_D^2/N - 1}} = \frac{\bar{D}}{\sqrt{s_D^2/N - 1}}$$

We test this statistic the same way as any *t* value, using alpha as the critical region with $N - 1$ degrees of freedom, where N is the number of people having matched scores.

Can the difference of means test be applied to the binomial case too?

Yes. As with the single-sample case, we can hypothesize a sampling distribution of the difference of proportions. We use the *z* value where the numerator will reduce to the difference of two sample proportions (the population proportions cancel out for the same reasons that the population means did; i.e., the difference is hypothesized to be zero). The denominator is the standard error of the difference of proportions. Thus, the z value is

$$z = \frac{P_{s_1} - P_{s_2}}{\sqrt{\dfrac{P_1 Q_1}{N_1} + \dfrac{P_2 Q_2}{N_2}}}$$

Note that under the null hypothesis of no difference between population proportions, the variances of the samples must be equal. Therefore, we need use only one formula.

Consider this example as a difference of proportions. A recent study shows that the proportion of females under 16 who have experienced sexual intercourse is .50 (i.e., $P = .5$). Twenty years ago, Kinsey's sample reported that the proportion was .30. Can we be sure (at the .05 level) that the incidence of intercourse is actually higher today, if each sample consisted of 100 people? Since N is large, we use z:

$$P_1 = .5 \qquad\qquad P_2 = .3$$

$$N_1 = 100 \qquad\qquad N_2 = 100$$

$$\sigma^2_{p_1} = \frac{P_1 Q_1}{N_1} = \frac{.5 \times .5}{100} = .0025 \qquad \sigma^2_{p_2} = \frac{P_2 Q_2}{N_2} = \frac{.3 \times .7}{100} = .0021$$

$$z = \frac{.5 - .3}{\sqrt{.0025 + .0021}} = \frac{.2}{\sqrt{.0046}} = \frac{.2}{.068} = 2.95$$

$$z_{.05} = 1.65 \qquad 2.95 > 1.65$$

We reject H_0: $P_1 = P_2$ and accept H_1: $P_1 > P_2$ since our computed z value exceeds the critical z. Therefore, we must conclude that the younger group does have a higher incidence of intercourse than Kinsey's sample.

Can we apply the difference-of-means test to a situation with more than two samples?

If we are testing a hypothesis using more than two samples, we could do a t test for all possible pairs of samples. This might prove unwieldy if the number of samples were quite large; for instance, 5 samples yields 10 t tests (remember the combination, 5 things taken 2 at a time?). One way to get around this is to do a t test for the largest difference of means. If they are not significantly different, chances are that other pairs would not be either.

However, it is possible to do a kind of multiple t test, testing the variation among means for all samples simultaneously. We are able to do this precisely because of the generalized relation between the t distribution and the F distribution. This simultaneous test of several sample means, called a one-way analysis of variance, is the subject we will next survey.

How Much Do You Remember?

New Words

Sampling distribution of the
   difference of means
Standard error of the difference
Chi square
F distribution
Homoscedasticity
Experimental and control group

New Symbols

$\sigma_{\bar{x}_1 - \bar{x}_2}$
χ^2
F

Did You Ever Wonder?

1. What is the difference between independent and dependent groups in the two-sample test?
2. Why is it better to assume the two population variances are equal?
3. Is it possible to pool the estimate of variance in computing the test for difference of proportions?

Want to Know More?

1. BLALOCK, H. M. *Social Statistics.* 2nd ed. New York: McGraw-Hill Book Company, 1972. Chapter 13.
2. GUILFORD, J. P. and B. FRUCHTER. *Fundamental Statistics in Psychology and Education.* 5th ed. New York: McGraw-Hill Book Company, 1973.
3. HAYS, W. L. *Statistics.* New York: Holt, Rinehart and Winston, Inc., 1963.
4. WILLIAMS, FREDERICK. *Reasoning with Statistics.* New York: Holt, Rinehart and Winston, Inc., 1968. Chapter 6.

Multiple samples

MUCH ADO
ABOUT MANY MEANS

14

How can the difference of more than two population means be tested?

We often test the difference of means measured in more than two nominal categories. To observe differences between blacks and whites without including orientals and others would be a misleading test of racial differences. Likewise, testing only the difference between Catholics and Protestants would offer a less than complete test of all religious differences. Political party, occupation, and a host of other nominal-level categories are subject to this same argument; therefore, we need a statistic to test the null hypothesis that multiple samples come from the same population. (H_0: $\mu_1 = \mu_2 = \mu_3$). Such a technique is called the **analysis of variance**. Analysis of variance is a direct extension of the use of the F test for the ratio of variances. The multiple-sample means test, therefore, is a statistical expansion of the two-sample test.

Think of this example: you are concerned over which brand of gasoline will give you the best mileage. Since you know that mass media advertising is mostly propaganda, you decide to conduct your own test. By keeping track of fill-ups, over several months, at each of three brand-name gas stations, you compute the mean mileage per gallon of gas from each of the three stations. The decision you must make is whether the average mileage per gallon for any station is sufficiently

145

The decision you must make is whether the average mileage per gallon for each of the three stations is sufficiently different to lead you to conclude that one or more of the brands is unlike the rest

different to lead you to conclude that one or more of the brands of gasoline is unlike the rest (i.e., comes from another population).

Unlike a two-sample test of means, we now have more than two categories, each measuring an interval-level trait. The strategy is to compare the variation among the means of the categories to the total variation within all the categories.

Why is it called the analysis of variance?

As mentioned earlier, analysis of variance is a direct extension of the F test for the ratio of variances. It is analogous to the t test, with these two exceptions: rather than the difference of two means for the numerator, we have the squared difference (i.e., variation) among more than two means. For the denominator, rather than utilizing an estimate of the standard error of the difference of means, we pool our estimate by adding each of the within-sample variances. Thus, we have

$$t = \frac{\bar{X}_1 - \bar{X}_2}{\sigma_{\bar{X}_1 - \bar{X}_2}} \quad \begin{matrix} \leftarrow \\ \\ \leftarrow \end{matrix} \quad \begin{matrix} \text{between sample} \\ \text{variation} \\ \text{within sample} \\ \text{variation} \end{matrix} \quad \begin{matrix} \rightarrow \\ \\ \rightarrow \end{matrix} \quad \frac{\hat{s}_B^2}{\hat{s}_W^2} = F$$

Analogy between t and F ratio

The numerator of the F ratio estimates the variance of the population means. The closer these sample means are in value, the smaller their variance and the more we support our hypothesis that they come from the same population (i.e., H_0: $\mu_1 = \mu_2 = \mu_3$). The larger the difference in sample means, the larger their variance, undermining the null hypothesis of no difference in the population. It is important to note that we measure this difference *relative* to the variation within the samples; that is, we measure the variation among the means in ratio to the variance within the samples. The greater the variance between the means relative to the variance within the samples, the more likely we are to reject the null hypothesis that the population means are equal; conversely, the smaller the variance between the means relative to the variance within the samples, the less likely we are to reject the null hypothesis. We are, therefore, **analyzing** these variances with the aid of the F statistic. In this case, it is the ratio of the variance among sample means to the variance within samples. We determine these two variances by partitioning the **sum of the squares**.

What does it mean to partition the sum of the squares?

Consider our example of the gas stations. Suppose we have data from our test (see table 14-1). Obviously, if we took every individual value in the three groups and added them together, dividing by the total N, we would have computed an overall mean of the three samples. You should satisfy yourself that this gives us exactly the same value as if we had taken the mean of each sample, and then computed the mean of those means:

$$\frac{\Sigma X_{ij}}{N} = \frac{\Sigma X_1 / n_1 + \Sigma X_2 / n_2 + \Sigma X_3 / n_3}{K}$$

where K = the number of samples

Therefore, we can consider an estimate of the total population variance

TABLE 14-1

		Stations			
		Gulfco	Texfield	Sea-Side	Total
Miles per gallon on fill-up	1.	12.2	11.7	12.9	
	2.	14.5	10.4	14.5	
	3.	11.4	8.8	12.1	
	4.	13.1	11.1	13.8	
	5.	9.8	12.5	11.2	
1. Sums		61	54.5	64.5	180
2. N (no. of cases)		5	5	5	15
3. Means		$\bar{X}_1 = 12.2$	$\bar{X}_2 = 10.9$	$\bar{X}_3 = 12.9$	$12 = \bar{X}_g$

to be the grand mean subtracted from each score, squared, summed up, and divided by the degrees of freedom $N - 1$:

$$\frac{\sum_i \sum_j (X_{ij} - \bar{X}_g)^2}{N - 1}.$$

Likewise, it should be clear that this is the same as taking the sum of the squares around each group mean and adding to that the sum of the squares of the group means around the grand mean (adding them and dividing by $N - 1$):

$$\frac{\sum_i \sum_j (X_{ij} - \bar{X}_j)^2 + n_i \sum_j (\bar{X}_j - \bar{X}_g)^2}{N - 1}$$

That is, these two variances are equivalent:

$$\frac{\sum_i \sum_j (X_{ij} - \bar{X}_g)^2}{N - 1} = \frac{\sum_i \sum_j (X_{ij} - \bar{X}_j)^2 + n_i \sum_j (\bar{X}_j - \bar{X}_g)^2}{N - 1}$$

Look at just the numerator for a moment. We see that $\sum\sum (X_{ij} - \bar{X}_g)^2$ $= \sum\sum (X_{ij} - \bar{X}_j)^2 + n\sum (\bar{X}_j - \bar{X}_g)^2$. The first term in this equation, which we call the **total sum of squares** (abbreviated Total SS), is the sum of the squared deviations from the overall mean. As we have just stated, this is equivalent to the sum of the squared deviations of the values from the group means, plus the group means squared deviation from the overall mean. We call the latter the **sum of the squares between** (the means) and the former the **sum of the squares within** (the groups). Symbolically it is indicated by between SS and within SS.

How is the between group sum of squares computed?

In its simplest form, it is the numerator of the variance among whatever number of means we have. By taking the overall mean from each group mean and squaring it, adding as many deviations of this value as there are cases in each group, then, we obtain the sum of the squares between. Take the data from our example above:

$$\bar{X}_1 = 12.2 \qquad \bar{X}_2 = 10.9 \qquad \bar{X}_3 = 12.9$$

$$n_1 = 5 \qquad n_2 = 5 \qquad n_3 = 5$$

$$\text{Between } SS = n_i \sum_j (\bar{X}_j - \bar{X}_g)^2$$

$$= 5(12.2 - 12)^2 + 5(10.9 - 12)^2 + 5(12.9 - 12)^2$$

$$= 5(.2)^2 + 5(-1.1)^2 + 5(.9)^2 = .2 + 6.05 + 4.05 = 10.3$$

How is the within group sum of squares computed?

Essentially, it is the numerator of the variance for each group, added together for *all* groups. We can compute this directly by subtracting the group mean from each score in the group, squaring the value and adding them, repeating for all scores in the group and summing across groups as in the example from our data:

$$\text{Within } SS = \sum_i \sum_j (X_{ij} - \bar{X}_j)^2$$

$$= (12.2 - 12.2)^2 + (14.5 - 12.2)^2 + (11.4 - 12.2)^2$$
$$+ (13.1 - 12.2)^2 + (9.8 - 12.2)^2 + (11.7 - 10.9)^2$$
$$+ (10.4 - 10.9)^2 + (8.8 - 10.9)^2 + (11.1 - 10.9)^2$$
$$+ (12.5 - 10.9)^2 + (12.9 - 12.9)^2 + (14.5 - 12.9)^2$$
$$+ (12.1 - 12.9)^2 + (13.8 - 12.9)^2 + (11.2 - 12.9)^2$$

$$= 0 + 5.29 + .64 + .81 + 5.76 + .64 + .25 + 4.41 + .04$$
$$+ 2.56 + 0 + 2.56 + .64 + .81 + 2.89$$

$$= 27.3$$

On the other hand, it is possible to calculate this indirectly as well, since we know that the total SS is equal to the between group SS plus the within group SS. That is, we determine the total SS by subtracting the group mean from each value and squaring, summing over all values. Subtracting the between SS from the total SS, we have computed the within SS. Again take our previous data:

$$\text{Within } SS + \text{ between } SS = \text{total } SS$$

$$\text{Within } SS = \text{total } SS - \text{between } SS$$

$$\Sigma\Sigma(X_{ij} - \bar{X}_j) = \Sigma\Sigma(X_{ij} - \bar{X}_g)^2 - n\Sigma(\bar{X}_j - \bar{X}_g)^2$$

$$27.3 = 37.6 - 10.3$$

How are mean square values derived from sums of squares?

Now that we have computed the between group SS, we should remind ourselves that initially we are trying to estimate the variance, or squared difference, among the population means. Since we are estimating the variance of population means from knowledge of our samples, we must use the best unbiased estimate of the variance. We learned, in chapter 12, that the best estimate of the variance comes when we divide the sum of squares by the degrees of freedom rather than the total N. In our example, we have three means of which two are allowed to vary before the third is determined. Thus, the estimate mean square is derived by dividing our sum of squares by the degrees of freedom. However, since we are using N to denote the number of cases in the samples, we usually express the degrees of freedom for the between variance as $K - 1$, where K is the number of group means. Hence:

$$MS_{between} = \frac{\text{between group } SS}{K - 1} = \frac{n_i \sum_j (\bar{X}_j - \bar{X}_g)^2}{K - 1}$$

It has the same interpretation as any sample variance, always defined as the sum of the squares per degree of freedom. Thus, we have the term average or **means squares between** (group means).

As you might guess, the computation of the MS within is analogous to the MS between. To estimate this variance, we must divide the within group SS by its proper degrees of freedom. All but one value is free to vary in each group, so taking N minus however many groups we have, we should establish the correct degrees of freedom. As stated above, the number of groups is usually denoted by the letter K (we use letters in the middle of the alphabet to count). Therefore, the proper degree of freedom is $N - K$. Our estimate of the variance is

$$MS_{within} = \frac{\text{within group } SS}{N - K} = \frac{\sum_i \sum_j (X_{ij} - \bar{X}_j)^2}{N - K}$$

Again, the **mean square within** has the same interpretation as any estimate of the variance with a certain degree of freedom. This is the average sum of squares per degree of freedom.

We should see the relationship between the two denominators or degrees of freedom in the following equality:

$$N - 1 = (N - K) + (K - 1) = N - K + K - 1 = N - 1$$

Hence $N - K + K - 1 = N - 1$ the degrees of freedom for the total sum of squares estimate of the population variance.

How is the F ratio interpreted in analysis of variance?

If you know from which group an object comes, your best guess of its value is the mean of that group. The extent to which the individual score is different from the mean of the group is the error in your guess. Therefore, think of the variance within the groups as an error variance; the error made in guessing, using the group mean to predict the value of the cases in each group. If all the group means are the same, this error variance (i.e., within group variance) is equal to the total variance in the samples. Thus, the error variance would be a good estimate of the population variance if the null hypothesis is true (that we are drawing samples from populations which have the same mean).

Likewise, if the null hypothesis is true, the variance between the sample group means is also a good estimate of the population variance, since the group means are merely linear combinations of the values in the group. Both variances are estimates of the same population variance under the null hypothesis. Therefore, it should seem reasonable that if they differ significantly we can have confidence the null hypothesis is not true. To put it another way, the degree to which the two estimates differ is the degree to which we support the alternative hypothesis (that the sample means come from different populations).

By putting these two estimates of variance in ratio to one another, we are saying that the larger the between mean variance relative to the within-group variance, the more we can be confident that the null hypothesis is false. The smaller the between-mean variance, the less confident we can be in rejecting the null hypothesis.

$$\begin{matrix} \text{Larger} \to \\ \text{or} \\ \text{Smaller} \to \end{matrix} \frac{MS_{between}}{MS_{within}} = \begin{matrix} \text{larger } F \\ \text{value} \end{matrix}$$

$$\begin{matrix} \text{Smaller} \to \\ \text{or} \\ \text{Larger} \to \end{matrix} \frac{MS_{between}}{MS_{within}} = \begin{matrix} \text{smaller} \\ F \text{ value} \end{matrix}$$

The relationship that these variances bear to each other is specified exactly by the probability distribution of F values.

What does the F distribution look like?

Since we are dealing with squared values, we can never have a negative number; therefore, all values of F must be positive. The

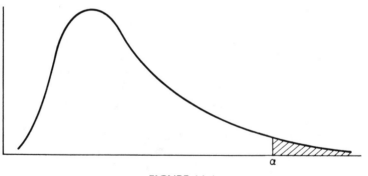

FIGURE 14-1

distribution reads, from left to right, zero to plus infinity (fig. 14-1). We can easily see that tests of significance using the F distribution are more likely to be one-tailed tests with the value specified by alpha and the appropriate degrees of freedom.

What conclusions can we draw from analysis of variance?

In computing the test statistic F from our example, we compiled the information in table 14-2.

TABLE 14-2

Variation	SS	df	MS	F
between groups	10.3	2 (K − 1)	5.15	2.26
within groups	27.3	12 (N − K)	2.28	

You will find it wise to put the information in similar table form, thereby avoiding the loss of the many scraps of paper with your carefully computed values on them.

After determining the F ratio, we refer to the table of F values (Appendix IV). Having decided beforehand on an alpha of .05 or .01, we proceed to look up the critical F value using the between group degrees of freedom to locate the proper *column* across the top, and the within group degrees of freedom to locate the proper *row* down the side. The intersection of this row and column contains the critical F value.

When we compare our computed F value to the value found in the table, the same decision rule applies. If our value exceeds that in the table, we reject the null hypothesis; if it does not, we do not reject the null hypothesis. The conclusion, therefore, is either (H_1) that the category sample means reflect separate population means, or (H_0) they are essentially one population from which our category means represent different possible samples.

In our gasoline example we have computed an F value of 2.28. The F table for alpha = .05 with 2 and 12 degrees of freedom indicates we must exceed a value of 3.88. Therefore, we cannot reject the null hypothesis; instead, we tentatively accept the assertion that there is no difference between gasolines.

Often we refer to an analysis of variance as testing the "effect" of the category on the interval-level measurement (e.g., race on I.Q.). If we reject the null hypothesis, we are saying that the attribute does have an effect on the interval-level variable (e.g., race affects I.Q.).

> *Can we test the effect of more than one variable*
> *on an interval trait?*

Yes. When we have more than one categorical variable, we often call our experiment a **factorial design**. *A factorial design is a test of the effect of more than one categorical attribute on an interval-level variable.*

Reverting to our example of the gas mileage, suppose more than one person in the family drives the car. It would seem natural to study the effect of the brand of gasoline *and* the person driving. Clearly these are both categorical attributes, but now we have increased our possible conclusions. We can test the effect of the gasoline, or the effect of the person driving, or the **interaction** of the two combined. When we have two such variables, we call this a two-way analysis of variance, or two-factorial design. The principle is the same as that of one-way analysis, except that we can test more than one effect. We usually summarize our data according to categories, such as presented in table 14-3.

TABLE 14-3

Member of Family	Station			Total
	Gulfco	*Texfield*	*Sea-Side*	*Total*
Father	12.2 14.5	11.7 10.4	12.9 14.5	$\Sigma_1 = 76.2$ $n = 6$ $\bar{X}_{1.} = 12.7$
Mother	11.4 13.1	8.8 11.1	12.1 13.8	$\Sigma X_{2j} = 70.3$ $n = 6$ $\bar{X}_{2.} = 11.72$
Son	9.8	12.5	11.2	$\Sigma X_{3j} = 33.5$ $n = 3$ $\bar{X}_{3.} = 11.17$
Total	$\Sigma X_{i1} = 61$ $n = 5$ $\bar{X}_{.1} = 12.2$	$\Sigma X_{i2} = 54.5$ $n = 5$ $\bar{X}_{.2} = 10.9$	$\Sigma X_{i3} = 64.5$ $n = 5$ $\bar{X}_{.3} = 12.9$	$\Sigma X_{ij} = 180$ $n = 15$ $\bar{X}_g = 12$

Since the table containing data for a two-way analysis of variance is presented in rows and columns, we call the effects of the two attributes the row effect or column effect, respectively. Essentially we are testing a one-way analysis of variance for the row and the column variables separately, and then again for their interaction. The result is three F ratios that we look up separately. In each case, we have used the row, column, and interaction MS as estimates of the population variance and compared them in ratio to the error term (MS within).

This idea may be extended to three or more factors. In addition, we can control for the effects of other interval variables, with a technique known as covariance analysis. Likewise, it is possible to set up elaborate experiments to randomize or sample certain variables, leading to use of techniques such as randomized block or latin square designs. It is beyond the scope of this book to explore these procedures, but the important point to remember is that, in each case, we are deriving an F ratio that is to be tested against a critical value in the F distribution so that a decision may be made about whether differences between population means actually exist.

> *Are there any tests of significance in which no interval (quantitative) measurement is assumed?*

In the multiple-sample test of means, and even in single-sample cases, we may violate certain assumptions about variances in the population, leaving us in some doubt about what effect this has on our decision rule. An even more common problem occurs where we cannot make the assumption of interval measurement. For example, can we really be sure that successive scores on a prejudice scale indicate equal intervals of intensity of feeling?

Techniques that sidestep these problems are often referred to as distribution-free statistics (as in the first case) or nonparametric statistics (as in the second case). In short, we do have specific procedures we can follow either to make inferences about nominal and ordinal data or to utilize when no assumptions can be made about the shape of the populations from which our interval-level samples were drawn. We will investigate these and other pertinent matters in the next chapter.

How Much Do You Remember?

New Words	*New Symbols*
Analysis of variance	SS
Sum of squares	MS
Total	$K - 1$
Between	$N - K$
Within	
Mean square	
Between	
Within	
Factorial design	
Effects	
Interaction	
Row	
Column	

Did You Ever Wonder?

1. Why is a comparison of more than two means a test of variance among them?
2. Why do we always divide the *SS* by the respective degrees of freedom in the *F* ratio?
3. Why are the values of an *F* distribution always positive?

Want to Know More?

1. EDWARDS, A. L. *Statistical Methods for the Behavioral Sciences.* New York: Holt, Rinehart and Winston, Inc., 1955.
2. FISHER, R. A. *The Design of Experiments.* 7th Ed. London: Oliver and Boyd, 1960.
3. KIRK, R. E. *Experimental Design: Procedures for Behavior Sciences.* Belmont, Calif.: Wadsworth Publishing Company, 1968. Chapter 2.
4. ROSCOE, J. T. *Fundamental Research Statistics for the Behavioral Sciences.* 2nd ed. New York: Holt, Rinehart and Winston, Inc., 1975. Chapters 35–36.

Nonparametric
Alternatives

IV

Measures
of differences

THE STURDY STATISTICS

15

What is the difference between nonparametric and distribution-free statistics?

Among applied researchers these two terms have become almost equivalent, although strictly speaking, they are not the same. A **nonparametric** *test makes no assumption about the value of the population parameters* (e.g., means and variances) from which the sample data have been compiled; *a* **distribution-free** *test does not make any assumptions about the shape of the population distribution.* In practice, however, nonparametric tests have come to apply to data in which we cannot assume quantitative (i.e., interval or ratio) measurement. Hence, these are tests that apply to nominal or ordinal classification. On the other hand, distribution-free implies that there may be interval or ratio measurement, but that we do not assume anything about the normality of the populations or sampling distributions.

Obviously these two definitions are not mutually exclusive. In fact, there exist distribution-free statistics that are parametric (e.g., sign test). Most of the discussion in this chapter, however, will treat these two terms synonymously and will take them to mean that we are dealing with techniques appropriate to nominal and ordinal measurement.

Are nonparametric tests less accurate than parametric tests?

"Accurate" is not the best word. You will recall in chapter 11 we introduced the concept of the power of a statistical test. It is the proportionate number of times that we correctly reject the null hypothesis. When referring to the power of a nonparametric test, in relation to the power of its parametric cousin, we coin the term "efficiency"; we may say that *the* **relative efficiency** *of a nonparametric test is the ratio of the power of the corresponding parametric test to the power of the nonparametric test*:

$$\text{Relative efficiency} = \frac{\text{Power of parametric test}}{\text{Power of nonparametric test}}$$

Because most nonparametric tests deal with small samples and/or differences in two samples, we often express relative efficiency as a ratio of the power of the *t* test to the power of the nonparametric test.

Though it is hard to generalize, a good ballpark figure of the efficiency of most nonparametric tests ranges from about .65 to 1.00, meaning we would correctly reject the null hypothesis using the nonparametric technique (given the fact that all the parametric assumptions are true, i.e., interval measurement and normality) 65 to 100 percent as often as we would correctly reject the null hypothesis using the parametric test. However, basing our evaluation on this premise (that the parametric assumptions are true) offers a misleading approach and conclusion. A better test is to see how both statistics fare when we know the parametric assumptions are *not* true. Under these circumstances, we find that the nonparametric tests are equally as powerful, indeed often much more powerful, than the corresponding parametric test.

What are the advantages of using nonparametric statistical tests?

In addition to a relaxation of requirements about the level of data, understanding the mathematical foundations is easier. Most of the test statistics and sampling distributions derive from the simple rules of counting and probability. Therefore, we receive much more of an intuitive return on our concentrated investment.

Likewise, since the mathematics are less involved (e.g., we use simple arithmetic operations of counting, ranking and subtracting) we save considerable time and effort in the computation, allowing greater emphasis to be placed on interpretation of the findings.

*Do we test hypotheses in the same way using
nonparametric tests?*

Yes. As we have stated, most nonparametric tests have a
sampling distribution from which a level of significance and critical region
can be determined. Once the test statistic is computed, the decision
is made in the usual manner by comparing the constructed value with
that found in the table. If the value exceeds the cutoff point (determined
by alpha), we reject the null hypothesis. If the value falls short of
the cutoff point, we do not reject the null hypothesis.

It is not the purpose of this book to derive all the sampling
distributions for each nonparametric technique. We will, however,
elaborate on their rationale, explaining the nonparametric techniques
analogous to the single-sample, two-sample, and multiple-sample parametric tests.

How are nonparametric tests applied to single samples?

Since we are not assuming anything about the population and
are dealing only with the sample data, the most persistent questions
concern the distribution of observations into nominal categories and
the cumulative distribution of observations into ordinal ranks.

We are essentially asking whether, in the absence of any influences,
the observed frequencies represent a random distribution. The most
popular statistic for the nominal case is an adaptation of the already
mentioned chi-square statistic. The statistic for rank data is named after
its authors: the Kolmogorov-Smirnov single-sample test, or more commonly, the Smirnov test. (The names are pronounced Coal·mo·gor·off
and Smear·noff and should not be confused with the vodka, although
the same country is famous for both.)

*What is the purpose of the chi-square test when used
on single-sample nominal-level data?*

Suppose you were in Las Vegas one weekend and happened
upon three slot machines standing prominently in a casino lobby. You
wanted to step up and insert a nickel, but better judgement told you
to wait and observe which machine, if any, was the most popular (this
is a sound strategy since popular machines usually pay more frequently).
Therefore, in the next hour you observe 60 people playing the various
machines. Now you must come to a decision; given the relative frequency
with which each machine was played, can you assume there really was

a popular one rather than merely random playing? Put more generally, is your empirical distribution so much different from a random distribution that we can assume something is causing a bias?

Problems of this sort in which we have k nominal categories (remember statisticians like to use i, j, k, to count), use **chi-square** to test the degree to which our observed distribution of frequencies approximates a random distribution. This test, called the "goodness of fit," tells us how well our empirical data fit the theoretically random allocation.

Remember that the chi-square distribution is a distribution of variances. Since the number or frequency, rather than the categories, is interval-level measurement, it should make sense that we could derive a measurement of variance in frequency by subtracting each expected frequency value from the observed, squaring it, and dividing by the frequency expected. Since this is a variance, we know it is distributed as chi-square. Hence, the chi-square distribution in the single-sample nomimal case can be defined as

$$\sum_{i=1}^{k} \frac{(f_o - f_e)^2}{f_e} = \chi^2$$

In the example of the slot machine, suppose we expressed the data as in table 15-1:

TABLE 15-1

Slot Machine	f_o Observed # of People	f_e Expected	$(f_o - f_e)$	$\dfrac{(f_o - f_e)^2}{f_e}$
No. 1	12	20	(-8)	$64/20$
No. 2	29	20	(9)	$81/20$
No. 3	19	20	(-1)	$1/20$
				7.3

$$\chi^2 = \Sigma \frac{(f_o - f_e)^2}{f_e} = 7.3$$

In table 15-1, chi-square is equal to 7.3. We compare this value with a critical value found in the chi-square table (see appendix IV). We choose the level of significance (across the top of the table), and determine the critical value by finding the appropriate degrees of freedom (down the side). In the single-sample case, the degrees of freedom are equal to the number of categories minus 1; that is, if we know the frequency in $k - 1$ of the categories and we know the total, the last category frequency is determined. Therefore, $k - 1$ values are free to vary.

$$\chi^2_{.05;2} = 5.99 \qquad 7.3 > 5.99$$

Our computed value (7.3) does exceed that which we looked up (5.99), and therefore, we reject the null hypothesis of randomness (i.e., there is a favorite machine).

A similar situation exists for the single-sample ordinal-level case; the difference is that we can cumulate the proportion in the rank categories. This involves the Kolmogorov-Smirnov test.

What is the Kolmogorov-Smirnov test used for?

The Kolmogorov-Smirnov single-sample test, or more simply the **Smirnov test**, is directly analogous to the chi-square test for single-sample nominal-level data, except for the advantage of our now being able to cumulate the frequencies in determining their deviation from randomness.

We might ponder, for example, whether the starting position in a track meet has an effect on winning; that is, whether the closer one is to the inside of the track, the more chance one has to win. Suppose we had compiled data on this phenomenon. Let F_0 stand for the cumulative distribution of proportions under the null hypothesis of randomness (F stands for the frequency, 0 is short for the null hypothesis), and S_n stands for the cumulative distribution of proportions in the sample of n trials. Then $|F_0(X) - S_n(X)|$ would be the absolute difference between what is random and what is observed. Call this by the letter D for difference. Thus, $D = \text{maximum } |F_0(X) - S_n(X)|$ (table 15-2).

TABLE 15-2

Position	Wins	Proportion (S_n)	Random (F_0)	$F_0(X) - S_n(X)$
1st	8	$8/20$	$4/20$	$4/20$
2nd	5	$13/20$	$8/20$	$5/20$*
3rd	3	$16/20$	$12/20$	$4/20$
4th	2	$18/20$	$16/20$	$2/20$
5th	2	$20/20$	$20/20$	0
	20	1	1	*$D = 0.25$

The sampling distribution for the differences is known and presented in appendix IV. We are looking for the maximum value of D, the difference which is greatest. This is a proportion which we can easily convert into a decimal. In the significance table, find the chosen alpha across the top and the number in the sample down the side. Compare our difference in decimals to the value in the table. The decision rule is the same as with other tests: reject if our value exceeds the critical level; do not reject if it does not. In our example, the maximum D = 5/20 or .25. At alpha = .05 and an N of 20, the critical value is .294. Our value does not exceed the critical value, and therefore we cannot reject the null hypothesis of randomness.

Can the chi-square and Smirnov tests be applied to two samples?

Yes. In fact they are both direct extensions of the single-sample case. For nominal-level data, suppose we have a cross-classification of attributes in the form of table 15-3. Notice that the f_e (shown enclosed in each cell) is found by multiplying the total in a given row by the total in the corresponding column and dividing by the total N. Since each cell's deviation from randomness can be treated separately, we can sum their squared deviations across all cells to produce a chi-square statistic. However, the degrees of freedom is expressed in a slightly different form. You will notice that we must know all but one value

TABLE 15-3

Sex	Political Party			Totals
	Republican	Democrat	Independent	
Male	$\frac{10.8\vert}{10}$	$\frac{13.2\vert}{14}$	$\frac{6\vert}{6}$	30
Female	$\frac{7.2\vert}{8}$	$\frac{8.8\vert}{8}$	$\frac{4\vert}{4}$	20
Totals	18	22	10	50

$$\chi^2 = \frac{(10-10.8)^2}{10.8} + \frac{(14-13.2)^2}{13.2} + \frac{(6-6)^2}{6} + \frac{(8-7.2)^2}{7.2}$$

$$+ \frac{(8-8.2)^2}{8.8} + \frac{(4-4)^2}{4}$$

$$= \frac{.64}{10.8} + \frac{.64}{13.2} - \frac{0}{6} + \frac{.64}{7.2} + \frac{.64}{8.8} + \frac{0}{4}$$

$$= .06 + .05 + .09 + .07 = .27$$

$$\text{df} = (r-1)(c-1) = 2 \qquad \chi^2_{2;.05} = 5.99$$

Since $.27 < 5.99$ do not reject H_0

in each of the rows and all but one value in each of the columns in order to specify the rest of the frequencies; as a result, the number of cells in the table which are free to take on any value is the number of rows minus 1 times the number of columns minus 1. Symbolically, $df = (r - 1)(c - 1)$. Thus, for our 2×3 table there are 2 degrees of freedom.

The decision is made in exactly the same manner as in the single-sample case. The null hypothesis is, of course, that the two nominal variables are independent of each other. In this example, the computed chi square is .27. The critical chi-square is 5.99. We cannot reject H_0, therefore, and tentatively assume sex and political affiliation are independent of one another.

In the ordinal two-sample case, the Smirnov test is identical with the one-sample version except that we are looking for the largest difference between two-sample cumulative distributions of proportions, rather than one-sample distribution and one theoretically random distribution. The rationale assumes that if the two samples are drawn from the same population, the two-sample cumulative distributions should be fairly close, whereas extreme differences would tend to support the alternative hypothesis that they are drawn from two different populations. Thus,

TABLE 15-4

		You	Cum %	Your Friend	Cum %	D	K_D	
	A	10	$10/25$	5	$5/25$	$5/25$	⑤	K_D at .05 with $N = 25$ is 9
	B	6	$16/25$	7	$12/25$	$4/25$	4	
Grades	C	6	$22/25$	8	$20/25$	$2/25$	2	$5 < 9$
	D	2	$24/25$	3	$23/25$	$1/25$	1	
	F	1	$25/25$	2	$25/25$	0	0	Therefore do not reject H_0
		25		25				

in the two-sample case D (represented by K_D to distinguish it from the single sample test) — maximum $|S_{n_1}(x) - S_{n_2}(x)|$. For instance, take a sample of your grades versus a friend's (table 15-4). Could you conclude these samples represent an overall difference of all grades? To test this hypothesis, we choose the appropriate alpha, decide if it is a one- or two-tailed test, and make the decision based on whether or not our computed value exceeds the K_D critical value. The larger the degree of difference, the greater the support for the alternative hypothesis that the two samples come from different populations.

K_D in the two-sample test indicates the numerator of the largest proportional difference, in this case 5. The critical value of K_D associated with an alpha equal to .05 and an N of 25 is 9 (see appendix IV). Since 5 is less than 9, we cannot reject H_0 and must tentatively conclude that there is no difference in overall grades.

Are there other nonparametric tests besides goodness of fit?

Yes, several. Often when dealing with two sets of ordered data, we can make use of the null hypothesis that the two samples should be randomly distributed when ranked as one distribution. Both the Wilcoxon rank-sum and the Mann-Whitney U tests measure randomization of this type. The appropriate statistic when the samples are dependent is the Wilcoxon matched-pairs signed-ranks test. In addition, the two-sample case of randomization of ranks can be straightforwardly extended to multiple samples. The Kruskal-Wallis test is such an extension.

What is the difference between the Wilcoxon rank-sum test and the Mann-Whitney U test for ordinal data?

Remember in grammar school, the teacher would match the girls against the boys in a race during physical education class? Didn't it always seem that the girls were faster than the boys at that age? Suppose you collected some data on the subject to see if this supposition were true. Let's say you rank order the speeds of all the children in one class without regard to sex. If there is really no difference between the two sexes, the ranks should have a random assortment of girls and boys among the ordering. One good index of this randomness is the *sum* of the values of the *ranks* in the group with larger N (either if N's are equal). This is the **Wilcoxon rank-sum** value T which can be thought of as the total sum.

Sex	Girl,	Boy,	Girl,	Girl,	Girl,	Girl,	Boy,	Boy,	Boy,	Boy
Rank	1	2	3	4	5	6	7	8	9	10

$$T = 2 + 7 + 8 + 9 + 10 = 36$$

The intuitive rationale is that the larger the sum of the ranks, the more likely it is that one group has a disproportionately large number of members in the last places. For this particular example, it turns out we would have needed a sum larger than 37 to reject the null hypothesis at alpha = .05. Thus, we cannot reject H_0 and assume that the boys and girls are not significantly different in speed.

The equivalent but more often used test for this type of problem is the **Mann-Whitney** U statistic. Here the difference is that we are interested in cumulating the number of ranks in one variable that precede the ranks of the other variable. In the example of races between boys and girls, we would want to know how many boys rank before each position containing a girl. The resulting sum is called U. The distribution would be

Sex	Girl,	Boy,	Girl,	Girl,	Girl,	Girl,	Boy,	Boy,	Boy,	Boy
Rank	1	2	3	4	5	6	7	8	9	10
No. of boys that beat girls in each rank	0		1	1	1	1				

$$U = 0 + 1 + 1 + 1 + 1 = 4$$

We make our decision based on the table representing the sampling distribution probabilities for U. Actually there are several tables (see appendix IV), each corresponding to the value of n_1, n_2, and U. In our example, $n_1 = 5$, $n_2 = 5$, which means that in order to reject at the .05 level, we would have to achieve a U statistic smaller than 3 (in our example of grammar school races $U = 4$). Here, as with T, we cannot reject the null hypothesis and therefore assume that the samples come from same population.

Actually, the rank-sum T and the U statistic can be shown to be related. The relative advantage is that the value of T is a bit easier to compute, while the table of the U statistic is a little more straightforward to interpret. Of course, both of these statistics apply only when the two groups are independent of each other; that is, when an observation in one group does not depend on observations in the other group.

What happens if the nonparametric samples are dependent?

Whenever we have one sample that we test on two occasions or two samples that have been matched, we must take their dependency into account and consider carefully the difference in scores. One nonparametric index of how different the two sets of scores are would be to rank the differences between each pair without regard to the sign. If the differences are random, we should expect just as many negative differences as positive differences (the sum of the ranks of the positive differences should be just about as large as the sum of the ranks of the negative differences). Summing the ranks with the *less frequent sign* yields a T value (Wilcoxon used T for his matched-sample statistic as well as for the rank sum). The T value can then be compared with the sampling distribution in (appendix IV). As in the parametric two-sample case, the probabilities vary with the number of cases, in this instance, with the number of differences.

Think of this example. You are an avid sports fan who wants to find out whether basketball teams have a tendency to do better or worse after the half-way point in a season. You collect data on seven teams chosen at random, with the results in table 15-5. For a two-tailed

TABLE 15-5

Team No.	Halfway Point		d	Rank of d	Rank with Less Frequent Sign
	Games Won Before	Games Won After			
1	35	40	−5	4	4
2	28	20	8	5	
3	18	15	3	2	
4	40	28	12	7	
5	33	35	−2	1	$\dfrac{1}{5} = T$
6	49	40	9	6	
7	22	18	4	3	

test at .05 with an N of 7, the table for T in appendix IV tells us we would have to achieve a sum of 2 or less in order to reject H_0. Since our value exceeds 2, we cannot reject the null hypothesis and must assume that there is no difference between games won before and after the break.

How can we use a nonparametric test for more than two groups?

Think of three groups of people with whom you interact; for example, your friends in school, your friends out of school, and older adults you know. Suppose you sampled each group giving each

TABLE 15-6

Score		
Friends in School	*Friends out of School*	*Adults*
83	60	97
86	75	63
92	73	88
98	80	90
72	85	93

an attitude scale measuring prejudice and obtained scores that are assumed to be ordinal data (table 15-6). If you rank these values as one distribution and sum the ranks of each group (let R stand for the sum of ranks), then their sums should be rather close if there is no significant difference

TABLE 15-7

	Rank	
Friends in School	Friends out of School	Adults
7	1	14
9	5	2
12	4	10
15	6	11
3	8	13
$R_1 = 46$	$R_2 = 24$	$R_3 = 50$

between groups (i.e., the ranks should distribute themselves randomly among the groups.) (table 15-7). It can be shown that if these orderings are randomly distributed, the **Kruskal-Wallis statistic:**

$$H = \frac{12}{N(N + 1)} \cdot \sum \frac{R_i^2}{n_i} - 3(N + 1)$$

is a good test of the difference in ranks among groups (for more on the mathematical reasoning see the references at the end of the chapter). Thus, by computing the value of H, we can test the null hypothesis that the ranks are randomly distributed among the three groups. In our example:

$$H = \frac{12}{N(N + 1)} \cdot \sum \frac{R_i^2}{n_i} - 3(N + 1)$$

$$= \frac{12}{15(15 + 1)} \cdot \sum \frac{(46)^2}{5} + \frac{(24)^2}{5} + \frac{(50)^2}{5} - 3(16)$$

$$= \frac{12}{240} [(423.2) + (115.2) + (500)] - 48 = 51.9 - 48 = 3.9$$

Looking up the critical value for H in appendix IV (the table for the Kruskal-Wallis test) we see that with 5, 5, 5 observations, we must exceed 5.67 to be significant at the .05 level. Our value does not exceed 5.67, therefore we cannot reject H_0, and assume that our friends have similar attitudes.

Does the size of the sample affect the nonparametric test?

In a way, yes. With small N's we can specify the sampling distributions exactly while when N's get larger, we must approximate the distributions. For example, the Kruskal-Wallis and Smirnov tests are approximated by the chi-square distribution when N's are large. The Mann-Whitney U and rank-sum and matched-pairs tests are approximately normal when N's are large. It is not the purpose of this book to derive the large-sample approximations, but rather to instill an intuitive feeling for the tests in general. Further interest should be directed to the selected references at the end of the chapter.

Are there other kinds of nonparametric tests?

Yes. Quite often, we want to test whether a significant relationship exists between two or more qualitative variables, rather than to test the differences between groups on one variable. Whenever we compute relationships between nominal or ordinal variables, we call them nonparametric measures of association. This is the topic of the next chapter.

How Much Do You Remember?

New Words	*New Symbols*
Nonparametric	D
Distribution-free	K_D
Relative efficiency	T
Smirnov test	U
Wilcoxon rank sum test	H
Mann-Whitney U test	
Wilcoxon matched pair test	
Kruskal-Wallis test	

Did You Ever Wonder?

1. What would be the meaning of "power" if we assumed that the parametric assumptions did not hold?
2. Why is chi square defined in terms of variance yet also used for testing randomness?
3. Why are the Wilcoxon rank-sum and Mann-Whitney U tests equivalent yet have different levels of significance?

Want to Know More?

1. BRADLEY, JAMES. *Distribution Free Statistics.* Englewood Cliffs, N.J.: Prentice-Hall, Inc., 1968.
2. MOSTELLER, F. and R. E. ROURKE. *Sturdy Statistics.* Reading, Mass.: Addison-Wesley Co., Inc., 1973.
3. PIERCE, ALBERT. *Fundamentals of Non-Parametric Statistics.* Belmont, Calif.: Dickerson Publishing Co., Inc., 1970.
4. SIEGEL, SIDNEY. *Nonparametric Statistics.* New York: McGraw-Hill Book Company, 1956.

Measures
of associations

HOW TO SAY A LOT
WHILE ASSUMING LITTLE

16

What are nonparametric measures of association?

We know that the correlation coefficient as a measure of association is functional in that it varies between ± 1.00 and that r^2 can be interpreted as a proportional reduction in the error of prediction. In nonparametric statistics, we will see that while the same notion of variance does not apply, we still typically deal in measures which have upper limits of 1.00, some of which also retain the proportional reduction in error interpretation. This is valuable in that it allows us to make use of unity as the measure of a perfect relationship, zero as indicating no relationship, and still retain the notion of reduction of error in prediction. *A nonparametric measure of association is the strength of the ability to predict the rank or category of one variable from another.*

One important rule to remember in computing qualitative measures of association: always use the statistic designed for the lower level of data. For example, if we want to measure the relationship between a nominal and an ordinal variable we use a nominal measure of association. Likewise, if an ordinal and interval variable are being related, we use an ordinal measure of association. This prevents faulty conclusions from violations in the measurement level.

How is correlation between ranks measured?

As mentioned in the historical note (appendix I), Spearman developed a method of measuring correlation between ranks soon after Pearson introduced the product-moment correlation coefficient. It was later redefined by Pearson himself; however, Spearman's name is still associated with it today. The symbol for **Spearman's correlation coefficient for ranks** is r_s. It uses the subscript s for Spearman so as not to confuse it with the product-moment correlation. (s is the preferred symbol as it cannot be confused with other measures; although ρ, and r_{rho} are sometimes seen).

The technique requires finding the difference between two ranks for each individual in the sample. Since the sum of these are, by definition, zero, we treat the squares of these deviations ΣD_i^2 as a measure of the discrepancy in the two ranks. Expressing this as a ratio to the total variance in the ranks (it can be shown that $N(N^2 - 1)/6$ is equivalent to the variance in the ranks), and taking that value from one (i.e., perfect prediction) we measure how *well* we predict from one rank to another. Thus, we have the definitional formula for the correlation between ranks:

$$r_s = 1 - \frac{6 \Sigma D_i^2}{N(N^2 - 1)}$$

This may look a bit exotic at first, but you must convince yourself that it is merely a rearrangement of the definition for Pearson's correlation coefficient when ranks rather than values are used. In fact, Pearson derived this directly from his formula for the product-moment correlation, the difference being that he substituted rank, rather than the value. (The interested reader should check the references at the end of this chapter for the derivation of this form.)

Meanwhile, as an example of the use of r_s, consider ten racing cars that compete together at two different tracks. Using the following

TABLE 16-1

Car No.	Race No. 1	Race No. 2	D_i	D_i^2
1	3	5	-2	4
2	4	4	0	0
3	7	6	1	1
4	9	9	0	0
5	10	7	3	9
6	6	3	3	9
7	2	2	0	0
8	1	1	0	0
9	5	8	-3	9
10	8	10	-2	4

$$r_s = 1 - \frac{6D_i^2}{N(N^2 - 1)}$$

$$r_s = 1 - \frac{6(36)}{10(99)}$$

$$= 1 - \frac{216}{990} = 1 - .22$$

$$= .78$$

results of their place finishes, we can compute r_s as a measure of how well these two ranks predict each other, that is, how equivalent the two rank orders are (table 16-1).

Observe that if the placings in the first race were the same as in the second, there would be no differences between the two and the sum of the differences ΣD_i^2 would be zero. Thus, the correlation would be perfect (1.00). Likewise, if one rank order were the reverse of the other, the scores would be maximally different. It can be shown very simply (try your own set of data) that the maximum squared differences is twice the variance in the ranks. Hence, the formula will reduce to $1 - 2 = -1$. These two facts imply that Spearman's rank-order coefficient varies between ± 1.00 (table 16-2).

TABLE 16-2

No. 1	No. 2	D_i	D_i^2	No. 1	No. 2	D_i	D_i^2
1	1	0	0	1	5	−4	16
2	2	0	0	2	4	−2	4
3	3	0	0	3	3	0	0
4	4	0	0	4	2	+2	4
5	5	0	0	5	1	+4	16
			$\Sigma D_i^2 = 0$				$\Sigma D_i^2 = 40$

$$r_s = 1 - \frac{6(0)}{5(24)} = +1.00 \qquad r_s = 1 - \frac{6(40)}{5(24)} = 1 - \frac{240}{120} = 1 - 2 = -1.00$$

One should be very careful not to interpret the *values* of Spearman's r_s as having ratio properties. Like Pearson's r, it cannot be said that an r_s of .6 is twice large as an r_s of .3. We can treat these values only as ordinal; that is, one is larger (or smaller) than another. We cannot meaningfully specify how much.

Is there an ordinal measure of association when the data are grouped?

We can think of grouping ordinal data in two ways. First, there could be an extreme number of ties in ranks, in which collapsing the ranks would form ordered classes. A better example of grouped data, however, is that presented in the form of a joint-frequency distribution where the categories are ordered classes (table 16-3). This is an example of grouped ordinal data. A measure of association quite easy to compute for these tables was developed by Goodman and Kruskal

TABLE 16-3   Cross-classification table for ordered data

	Upper	Middle	Lower	Total
Upper	5	3	2	10
Middle	10	8	2	20
Lower	5	9	6	20
Total	20	20	10	50

in 1954. It is called **gamma** and is symbolized by the Greek γ.

How is gamma computed?

If we are computing Spearman's r_s and have more than a few ties in the two respective ranks, the resulting coefficient will be artificially inflated. We need a measure free of ties within ranks. One way to solve this problem might be to ignore ties altogether. By collapsing rankings into ordered classes, we can compute the number of individuals measured on the first variable which can be paired with those individuals having a higher rank on the second variable. This can be thought of as the number of similar pairs or **concordance** (concordance means agreement). Likewise, if we compute the number of individuals measured on the first variable who can be paired with individuals ranked lower on the second variable, we can think of this as the number of dissimilar pairs or **discordance** (i.e., disagreement). The number of similar pairs plus the number of dissimilar pairs should equal the total number of possible pairs not tied. If the difference between similar and dissimilar pairs (i.e., similar minus dissimilar) is expressed in ratio to the total number of united pairs (i.e., similar plus dissimilar), we would have a measure of association reflecting the degree to which there is an increasing or decreasing trend from one variable to the other. Hence, we define gamma as

$$\gamma = \frac{C - D}{C + D}$$

where C is the concordance and D is the discordance.

Think of the analogy of your spelling tests in the third and fourth grades. If you could do only better or worse from one grade to the next, never remaining the same, then the number of improved spelling tests minus the number that were worse divided by the total number of tests should be a good index of how well your skill is improving. If all tests in the fourth grade were improvements, then the ratio would be 1.00, meaning perfect improvement or increasing ability. If all the tests were worse, the ratio would be -1.00, meaning perfect deterioration of ability. Thus, the measure of change in ability takes values from 1.00 to -1.00.

Gamma is interpreted the same way when frequency is in ordered classes. Take the example of a father's social class versus a son's social class (we often use this relationship as a measure of mobility) (table 16-4). If we order the classes from the upper left-hand corner down

TABLE 16-4   Frequency of father by son's social class

Father's Social Class	Son's Social Class			
	Lower	Middle	Upper	
Lower	10	3	7	20
Middle	4	3	3	10
Upper	6	10	4	10
	20	16	14	40

$$C = 10(3 + 3 + 10 + 4) + 3(3 + 4) + 4(10 + 4) + 3(4) = 289$$

$$D = 6(3 + 3 + 3 + 7) + 10(3 + 7) + 4(3 + 7) + 3(7) = 257$$

$$\gamma = \frac{C - D}{C + D} = \frac{289 - 257}{289 + 257} = \frac{32}{546} = .06$$

and across, we see that the number of similar pairs is equal to the number in each cell times the sum of the frequency in all cells higher on the other variable (the sum of all cells down and to the right). Likewise, the number of dissimilar pairs is equal to the number in each cell times the sum of the numbers in cells lower on the other variable (cells up and to the right). The difference of these two values divided by their sum is gamma. It varies between ± 1.00 and can be interpreted as the degree to which values on one set of ordered classes are associated with higher or lower values on the second set of ordered classes.

A special case of gamma is the 2×2 table consisting of ordered classes. The computation is exactly the same as with gamma, but since

TABLE 16-5

$$\text{Yule's } Q = \frac{AD - BC}{AD + BC} \quad \text{where}$$

	Low	High	
Low	A	B	$A + B$
High	C	D	$C + D$
	$A + C$	$B + D$	$A + B + C + D$

this special case was developed first, by a fellow named Yule (and called **Yule's Q**), the statistic has survived until today and therefore deserves mentioning (table 16-5). AD can be thought of as the number of similar pairs and BC is the number of dissimilar pairs. Therefore, $AD - BC = C - D$. Thus, the statistic yields exactly the same value as gamma and is interpreted in the same way.

Can measures of association be computed for nominal-level measurement?

Yes! The 2×2 table is particularly well suited for computing exact probabilities. One measure, developed by R. A. Fisher and known as **Fisher's exact test**, is well suited for small sample problems when any cell size is less than five.

Referring to the 2×2 table above (table 16-5) labeled A, B, C, D we should see that the number of ways of taking A things from $A + C$ things is $\begin{pmatrix} A + C \\ A \end{pmatrix}$. Likewise, the number of ways of taking B things from $B + D$ things is $\begin{pmatrix} B + D \\ B \end{pmatrix}$. Since only two cells are free to vary before the rest are determined, expressing these as a proportion of the number of ways of taking $A + B$ things from the total $\begin{pmatrix} N \\ A + B \end{pmatrix}$, is the exact probability of finding a distribution with these frequencies. Remembering what we learned about the chapter on probability, it is easy to reduce this proportion to

$$\frac{\begin{pmatrix} A + C \\ A \end{pmatrix}\begin{pmatrix} B + D \\ B \end{pmatrix}}{\begin{pmatrix} N \\ A + B \end{pmatrix}} = \frac{\frac{(A + C)!}{A!C!} \cdot \frac{(B + D)!}{B!D!}}{\frac{N!}{(A + B)!(C + D)!}}$$

$$= \frac{(A + B)!(A + C)!(B + D)!(C + D)!}{N!A!B!C!D!}$$

This qualifies as an apt measure of association because like all probabilities, it has an upper limit of 1.00 and a lower limit of 0.00.

Take the example of a survey made of male and female friends to see who does or does not own a car. You might suspect that males are more likely to own cars, so you collect data and calculate Fisher's

TABLE 16-6

	Male	Female	
Car	5	3	8
No car	0	2	2
	5	5	10

$$P = \frac{8!5!5!2!}{10!5!3!0!2!} = \frac{20}{90}$$
$$= .22$$

exact test to see (table 16-6). In other words, the probability of getting a distribution of frequencies such as this, under the null hypothesis of no relationship, is .22. If no cells contain zero, we must compute the probability for each case where the smallest cell size is closer to zero. We do this by constructing a new table, rearranging the frequencies to make them more extreme (i.e., the smallest closer to zero) and computing the test for each table (table 16-7). The probability of a table as extreme in frequency is just the sum of the probabilities computed for each table.

TABLE 16-7

3	3	6
2	2	4
5	5	

4	3	7
1	2	3
5	5	

5	3	8
0	2	2
5	5	

Are other measures used in the 2 × 2 Table besides Fisher's exact test?

Yes. In the chapter on nonparametric tests of differences (chap. 15), we introduced the chi-square statistic for measuring deviation from randomness in a frequency distribution of nominal categories. This had a probability interpretation when using the chi-square tables (see chap. 15). But we know that chi square is directly related to the number of cases. In particular we can show that the upper limit of chi square in the 2 × 2 case is N. Therefore, if we divide χ^2 by N, we should have a measure which varies between zero and one, in which zero shows no relationship and one shows perfect relationship. Such a measure is called **phi-squared** and is symbolized by the Greek letter ϕ. Thus:

$$\phi^2 = \frac{\chi^2}{N} \quad \text{where} \quad \chi^2 = \Sigma \frac{(f_o - f_e)^2}{f_e}$$

We can test the significance of this statistic by merely testing the significance of chi square.

Can ϕ^2 be used for tables larger than 2×2?

Yes. The problem is that chi square will no longer reach an upper limit of N when we compute a measure for a table larger than 2×2. It can be shown that the upper limit for larger than 2×2 is actually N times the smaller number of either rows or columns, minus one, that is, $L - 1$. (Think of L as standing for the smallest number of rows or columns.) Thus, in the 2×2 table we have $(L - 1)$ times $N = (2 - 1) \times N = N$. In general:

$$\phi'^2 = \frac{\chi^2}{N(L-1)} \qquad \phi' = \sqrt{\frac{\chi^2}{N(L-1)}}$$

where the new value is called phi prime. It is also called **Cramer's V** after its developer Cramer. This value will always have an upper limit of unity and a lower limit of zero. It is interpreted as the degree to which one variable is independent from the other. Its significance is found by testing the chi-square value for significance.

Take for example a cross-cultural survey done to see which men prefer which color hair (table 16-8). If these hypothetical data were

TABLE 16-8

	America	Europe	Orient	Total
Blonde	3.3⌋ 4	3.3⌋ 2	3.3⌋ 4	10
Brunette	3.3⌋ 3	3.3⌋ 5	3.3⌋ 2	10
Redhead	3.3⌋ 3	3.3⌋ 3	3.3⌋ 4	10
	10	10	10	30

$$\phi' = V = \sqrt{\frac{\chi^2}{2N}}$$

$$\chi^2 = \frac{(.7)^2}{3.3} + \frac{(1.3)^2}{3.3} + \frac{(.7)^2}{3.3} + \frac{(.3)^2}{3.3} + \frac{(1.7)^2}{3.3} + \frac{(1.3)^2}{3.3} + \frac{(.3)^2}{3.3} + \frac{(.3)^2}{3.3} + \frac{(.7)^2}{3.3}$$

$$.15 + .51 + .15 + .028 + .876 + .51 + .028 + .028 + .15 = 2.43$$

$$V = \sqrt{\frac{2.43}{60}} = .20$$

true, we could state that with a given probability, that on a scale from .00 to 1.00, these variables are related .20.

Is lambda different from the phi coefficient?

Phi is a measure of the strength of dependence in two sets of nominal categories. It is directly related to chi square and therefore, we can test its significance using χ^2. However, when the number of categories becomes quite large, phi has little commonsense meaning. Fortunately, there is a measure that retains a commonsense meaning and is based on error in prediction of the frequency in one variable given the category of the other. This statistic is called **lambda** and is symbolized by the Greek letter λ of the same name. It was developed by the same people and in the same year as gamma (Goodman and Kruskal, 1954), and has the same proportional reduction in error interpretation. That is, we can specify the reduction in the error we will make in predicting the values of one variable given information about the other. Take the data in table 16-9. If I didn't know anything about

TABLE 16-9

	A_1	A_2	A_3	A_4	Total
B_1	3	④	1	2	10
B_2	⑥	2	3	⑤	[16]
B_3	2	3	⑥	3	14
Total	11	9	10	10	30

variable A, my best guess of the values of B would be the largest marginal total for B (16). That is, I would be wrong less than for any other choice in predicting B. However, if I know what category on variable A that I am referring to, I can improve my guess of B considerably by choosing that category of B with the highest frequency in the column of A (for A_1, it would be 6; for A_2, 4; etc.). How much better I predict by knowing the categories of the other variable expressed as a proportion of the total possible improvement is precisely how we define lambda. Thus:

$$\lambda_B = \frac{\Sigma_{max} f_{B_i} - max f_B}{N - max f_B}$$

In the example above, $\lambda_B = (6 + 4 + 6 + 5 - 16)/30 - 16 = 5/14 = .36$. This is a measure of the ability of A to predict B. Obviously, we could determine the ability of B to predict A (called lambda$_A$) by switching the marginals, and cell-frequency guesses, from columns to rows. These are considered asymmetric measures in the sense that they predict in one direction. A symmetric or bidirectional prediction measure could be derived by merely adding the two lambdas. Thus:

$$\lambda_{AB} = \frac{\Sigma \max f_{B_i} + \Sigma \max f_{A_i} - \max f_B - \max f_A}{2N - \max f_B - max f_A}$$

Why do we call gamma and lambda "proportional reduction in error" measures?

For gamma, the sense of "error" in prediction is the "reverse" ordering of one variable given a particular value of the other. For lambda, the error is a sense of misclassification of one variable given a particular other. Error in gamma is the deviation from .00. That is, when excluding ties, we expect the values on the second variable to go up half of the time and down half of the time if the variables are independent. For lambda, error is in deviation from the modal category used for predicting. Both are expressed in proportion to the total number of guesses which are not in error. Thus, the reduction in error is the extent to which we can eliminate these misjudgments in prediction.

How Much Do You Remember?

New Words	New Symbols
Rank Order Correlation	r_s
Gamma	γ
Concordance	Q
Discordance	ϕ
Yule's Q	V
Fisher's Exact Test	λ
Phi	
Cramer's V	
Lamdba	

Did You Ever Wonder?

1. What would happen to r_s if too many ties appeared in the ranks?
2. Would gamma or phi be greater for the same data?
3. What is the advantage of using Fisher's Exact Test in a 2 × 2 table?

Want to Know More?

1. BLALOCK, H. M. *Social Statistics.* 2nd ed. New York: McGraw-Hill Book Company, 1968, Chapters 17 and 18.
2. COSTNER, H. L. "Criteria for Measures of Association," *American Sociological Review,* vol. 30 (1965): 341–353.
3. GOODMAN, LEO, and KRUSKAL, WILLIAM. "Measures of Association for Cross Classification," parts 1–3, *Journal of the American Statistical Association,* vol. 49 (1954): 732–764; vol. 54 (1959): 123–163; vol. 58 (1963): 310–364.
4. MUELLER, J. H., K. SCHUESSLER, and H. L. COSTNER. *Statistical Reasoning in Sociology.* 2nd ed. Boston: Houghton Mifflin Company, 1970.

Epilogue

I'm often reminded of the story about Karl Pearson, the great statistician, who was asked the question, "What is the first thing you remember?" He recounted that it was sitting in a high chair sucking his thumb when someone told him to stop sucking, stating that unless he did so, the thumb would wither away. Upon hearing this, young Karl put his two thumbs together and looked at them for a long time. "They look alike to me," he said to himself. "I wonder if she could be lying to me?"

Such an appeal to empirical evidence for more objective proof has long been the foundation for systematic social inquiry. In the light of the myriad of techniques that have developed since the time of Pearson, however, a related problem has emerged. The problem is one of using faulty assumptions, manipulated interpretations, misleading descriptions, and invalid conclusions in presenting a case for empirical evidence. In short, it has reached the point where selective presentation means that practically any point can be proved using statistics. The result is that people have become alienated from the subject, and often ask the question, Why bother studying it?

What must be remembered, however, is that statistics is like any other collection of assertions on how to interpret the world. Statistics are run rampant by those who mislead, misinform, and generally manipulate statements to influence people who do not have the knowledge or access to understand the facts. With the diffusion of mass media and the use of high-speed computers, this manipulation process has become even more accelerated. The logical conclusion, therefore, is to abandon any confidence in statistical assertions. This conclusion serves a dual purpose for the layman. First, it guards against falling prey to the "evil people" who lurk behind computers and calculators and the individuals who distort this mass of mental mumbo-jumbo. But more important, it provides a perfect rationale for not having to rekindle the anxiety that formal exposure to a statistics class generates.

For those who would cast out statistics because of vulnerability to its manipulation, think of an automobile. Because it may infrequently fail to operate for you, do we cease to use it? Do we stop eating because a meal has given us a case of stomach flu? Do we abandon government because an elected official misuses power? Of course not! We try to find out why something goes wrong and use that knowledge to correct the situation. Likewise, the misuse of statistics will not be cured by ignoring the subject. Rather, the diligent student who acquires knowledge about statistics is assured immunity from its abuse and misuse.

For those who would shy from the study of statistics because of mental anguish over mathematics, think of it as you would cooking a meal or fixing a car. We endure mental and physical discomfort in acquiring these skills because there is a reward for doing so. The finished meal provides nourishment and gives satisfaction from those who compliment it. Fixing a car oneself saves on costly automotive repairs. Accomplishing either task involves a sense of gratification for having completed the job. The same thing is true of statistics. By associating our study with the utility of using the methods, we will derive satisfaction in having completed the task.

Appendixes

Historical notes

I

Descriptive statistics

Although nobody really knows when numbers were first used to describe collections of objects, it's clear that they have been around since the dawn of recorded history. Economies with money as the medium of exchange were well known before the time of Aristotle (third century B.C.).

At least one source has traced the use of statistics as a descriptive device to "political arithmetic." The bosom of this concern was London, where John Graunt (1620–1674), a merchant and ex-army officer, wrote an extraordinary book in 1662 entitled *Natural and Political Observations upon the Bills of Mortality*. It was originally published to keep track of the plagues that were consuming much of the population in London at the time, but later became a source for measuring population growth. This work stands out as the first definitive use of **frequency** counts in a population and served as a statistical summary for periodic newspaper reports of births and deaths during the first half of the 1600's.

Graunt's tables of mortality did not go without notice by one friend in the academic world, a professor at Oxford by the name of William Petty (1623–1687). It was he who eventually coined the term *Political Arithmetic*. In his book by the same name, written in 1671 but not published until 1690, he firmly advanced the thesis that not only were

187

vital statistics important for a government to possess but information on the inhabitants was desirable as well. In this work, we find extensive use of frequency counts, such as the number of houses in London; however, Petty also made considerable use of **proportions** and **percentages** in enumerating the population.

Petty's writings caught on in the academic world and were passed around the Royal Society in London, a prestigious intellectual organization, which Petty helped found. During this flurry of exchange on the subject, Sir Edmund Halley (1656-1742), the brilliant astronomer who discovered the comet bearing his name, gathered data that became the major part of his tables of mortality, published in 1691, and entitled "An Estimate of the Degrees of the Mortality of Mankind Drawn from Curious Tables of the Births and Funerals at the City of Breslaw." Although many argued with Halley's assumptions and the validity of his tables, very little was added to the subject for the next 50 years, and in 1756 a life insurance company, the Equitable, was founded on Halley's methods.

Meanwhile, in Germany in 1761, a work called the *Divine Order* was published by J. P. Susmilch (1707-1767). In this work we have the first major concern over the concept of a **ratio**—in this case, the sex ratio. Susmilch's conclusion was that the proportion of men to the proportion of women remained constant because of a divine order imposed on populations. Its importance derives from the extreme influence it had on contemporaries in reinforcing the idea that research on population questions was a most important aspect of politics.

By the turn of the century (1798), an English preacher named Thomas Malthus had written a book entitled *Essay on the Principle of Population*, in which he firmly entrenched the idea of using a **rate**—in this case the population growth rate. It was Malthus's contention that populations grow geometrically, while food supplies are replaced at a constant rate. Therefore, restraint is needed to check widespread famine.

In the 1780's graphic representation of statistical data began to appear. A German, A. F. W. Crome (1753-1833), seems to have been the first to make use of geometrical graphs in representing population figures for most of the European states in 1785. Concurrently in England, in 1786, William Playfair (1759-1823) published the first use of a **bar graph** to explain the balance of trade for the British government.

The interest that these and other writers aroused around the world moved many countries to establish agencies to conduct the census. The Continental Congress in the United States provided for a regular census and the first was carried out in 1790. France established the Bureau of General Statistics in 1800 and proceeded to conduct the first census at that time. England, after 30 years of arguing, took the first census in 1801.

By the year 1830, enthusiasm over political arithmetic began to flourish. Several statistical institutions were founded and many statistical societies were formed. In England, in 1833, a statistical department was included in the Board of Trade. In 1837 a civil registration of vital statistics was established. In Germany, the Tariff Union was founded in 1833 to oversee their census. In 1839, the American Statistical Association was established in Boston, and in London the Statistical Society was formed in 1834 on the suggestion of Jacques Quetelet (1796-1874), head of the Belgian Statistical Commission.

Although the use of measures of central tendencies had been around since the time of Pythagoras, it was not until Quetelet's breakdown of the first Belgian census (1846) that we see a concerted effort to establish descriptions of the average man. While having to wait until 1893 for Karl Pearson (1857-1936) to popularize the notion of **standard deviation**, Quetelet's efforts had a direct effect on the study of educational testing and have earned him the title of "father of modern descriptive statistics."

Many contributors have influenced and guided research in describing the relationship between two variables, but the first formal presentation of the concept of a statistical relationship is found in the work of Francis Galton (1822-1911). His book, *Natural Inheritance*, published in 1899, contains the first statement on the use of one variable to predict another. Galton's work chiefly derives from the interest generated by two people. One was Charles Darwin, a cousin of Galton, who in 1859 published the *Origin of Species*. Such a profound effect did this book have on Galton's thinking that he devoted much of the rest of his life to the study of inheritance. However, it was not until early 1873 that Galton's work focused on prediction as a tool for the analysis of heredity. At that time a man by the name of Bowditch was involved in a study on the growth of children sponsored by the Massachusetts Board of Health. In it, records of bodily measurements, age, occupation of parents, nationality, and birthplace were compiled for some 24,000 students. One of Bowditch's techniques for analyzing this mass of data was to chart the joint occurrence of pairs of these variables, such as height and weight, by putting the frequency of occurrence of different heights for specific increasing values of weight. Galton seized on this idea for use in heredity research by charting the size of parent and offspring sweet pea seeds. Later efforts led to a study of the stature of fathers and sons. One of his curious findings showed that the stature of adult offspring on the whole was not as extreme and showed less variability than the stature of the parents. The original term Galton applied to this phenomenon was "reversion" (and thus the symbol r), which refers to the tendency for the mean offspring type to "revert" to the average parental type. Galton later called this occurrence **regression**, and the symbol r was used subsequently to describe the more general phenomenon of **correlation**.

The first use of the term correlation appears in a paper Galton gave in 1888 on "Co-relations and Their Measurement." In it, he makes his classic statement, "Two variable organs are said to be co-related when the variation of one is accompanied on the average by more or less variation of the other, and in the same direction . . . but the index of co-relation, which is what I call regression, is different (for different directions of prediction.)"

Several important contributions appeared just before the publication of *Natural Inheritance.* In 1895, the great statistician and friend of Galton, Karl Pearson (1857–1936), published his paper "Regression, Heredity and Panmixia," in which he derived the present-day formula for the correlation coefficient and the test statistic for its sampling distribution. Two years later, George Yule, a student of Pearson, developed the general formula for the regression coefficient b by the use of least-squares techniques and extended this notion to **multiple regression**. In this paper, we see the first use of R to indicate **multiple correlation**, as well as what Pearson later called **partial correlation**.

The first International Statistical Congress was to meet in Brussels in 1853. Within several years of this first meeting, we begin to see a meshing of descriptive statistics with the other tributary which was simultaneously yet separately evolving, that is, the development of the theory of probability.

Inferential statistics

Traces of probability statements can be found as far back as the walls of Egyptian pyramids, which show the games of rulers. The first written work on the subject of probability, however, appeared in a commentary on Dante's *Divine Comedy* in 1477, where reference was made to the probability of various outcomes in the throw of three dice.

This is typical of the beginnings of the history of probability. Early work started at the gaming table and was perpetuated by those fascinated with gambling. It is fitting, therefore, that history should record Jerome Cardan (Geronimo Cardano in Italian, 1501–1576) as the pioneer in the field of probability. The son of a lawyer and proficient in the fields of mathematics, physics, astronomy, and medicine, he was also an avid gambler. One of his writings, which roughly translated means "Book on Games of Chance," is really a handbook for gambling. Among other things, it contains sections on **relative frequency** and additive properties of **probabilities** and the idea of expectation. Cardan led a rather traumatic life and when his son was executed for wife killing, his colleague was poisoned, and he was arrested and charged with impiety (for, among other things, casting a horoscope of Jesus Christ), history writes that he became very bitter and senile. The fact is recorded that Cardan even predicted the day of his death. Not wishing to fall from grace at his final moment, he starved himself for three weeks in order to fulfill the prophecy.

Although hinted at by other writers in the interim, formal work on probability did not proceed again until a century later when Blaise Pascal (1623–1662) and Pierre de Fermat (1608–1665) began their famous correspondence in 1654 on the problem of two players of equal skill competing against each other for points by betting certain amounts of money. The various solutions for this game of chance led directly to the notions of **permutations** and **combinations**. Pascal's triangle mentioned in the text is a direct result of this correspondence.

Since Cardan's book was not published until 1663, and the Pascal-Fermat correspondence was never publicly circulated, it remained for Christian Huygens (1629–1695) to publish the first scientific treatise on probability in 1657. Huygens was an astronomer and physicist, but shared with Cardan an interest in the enumeration of mathematical probabilities in games of chance. The book is especially noted for the treatment of **mathematical expectations** (i.e., average outcome in the long run).

One of Huygens' students, Gottfrid von Leibniz (1646–1716), who became largely responsible for the invention of calculus, also contributed to the rules of probability and mathematics. Following the work of Pascal, he introduced the present-day notation for **combinations**: $C(n,r)$. It was Leibniz who also popularized the notation for **addition, subtraction**, and **equality** $(+, -, =)$.

Another follower of Huygens was from the most famous family of mathematicians the world has ever known. Jacob Bernoulli (1654–1705), the oldest in a family of nine famous mathematicians, began studying mathematics as a young boy, against his father's wishes. However, with Leibniz's encouragement and after a study of Huygens' work, he published *Ars Conjectandi*. This is the first book written solely on the subject of probability and certainly the first dealing with it as a separate branch of mathematics. The first section of this book is a reprint of the famous work of Huygens. Unfortunately, Jacob died before completely finishing the manuscript and it was up to Nicolas Bernoulli II, his nephew, to edit it posthumously for publication in 1713. The work is most noted for its exhaustive treatment of the two outcome sampling distribution (which later became known as Bernoulli trials), and expansion of terms in the **binomial probability distribution**.

One man who maintained contact with the Bernoulli family was Pierre Raymond de Montmort (1678-1719). His book, *Essai d'Analyse sur les Jeux de Hazards* (1708), shows an extensive treatment and extension of Bernoulli's work. Again, one can see the influence of games of chance and gambling providing the motivation. The first part concerns the theory of combinations; the second part discusses games of chance that depend on cards; the third part treats games that depend on dice; and the fourth part consists of solutions to various problems of probability being passed around at that time. Intended for a mathematical audience, it enjoyed wide appeal and did much to spread the study of probability.

Perhaps the first definitive statement made on the normal probability distribution or normal curve was by Abraham De Moivre (1667-1754). De Moivre never had a formal teaching position and spent most of his time tutoring. He was, however, greatly influenced by Bernoulli and Montmort. His first work, entitled *Doctrine of Chances* (1718), shows the influence of gambling and games of chance, but it also lays the groundwork of the normal curve by discussing at length the relationship between terms in a binomial expansion. In fact, it is here that we first observe the popular notation for the binomial expression: $(a + b)^N$. However, it was in a seven-page paper given to some friends on November 12, 1733, that he first expressed the relation between the binomial expansion and what he called the "curve of error." Understandably, this date is generally recognized as the origin of the **normal probability distribution**.

Since De Moivre could not guess the import of his discovery for the science of statistics, it took 100 years and the work of P. S. Laplace (1749-1827) to consolidate works of many writers into the single most authoritative work on the subject of probability, entitled *Theorie analytique des probabilities*, published in 1812. It contains among other things, the notion of the mean of the curve of errors, and an elaboration of the least-squares property of the mean, a property that paved the way for the understanding of the standard deviation and the central limit theorem.

Carl Friedrich Gauss (1777-1855), probably one of the three top mathematicians the world has ever known, followed Laplace in his work on the normal curve. So important was his influence that the normal curve is sometimes called the Gaussian curve. Gauss's earliest work was as an astronomer. In charting the orbits of planets, he became quite interested in deviations of the planets from their predicted paths.

This quite naturally led him to the study of the curve of errors and the method of least squares. By applying the calculus of probability to the notions of De Moivre and Laplace, Gauss extended the binomial interpretation of the normal curve of errors to continuous distributions. This extension, later refined, became known as the **Central Limit Theorem**.

As mentioned earlier, it was Quetelat who first realized the importance of the use of mathematical statistics in studying empirical events. His work greatly influenced the use of the normal curve in measurement of physical and mental traits and is reflected in his paper to the British statistical association (1841) on scientific observation. However, Francis Galton (1822-1911), a non-mathematician, firmly established the use of the normal curve of error in advocating its use for educational institutions in assigning grades and keeping educational records.

The actual unit of error used was not standardized until 1893, when Karl Pearson (1857-1936) coined the term **standard deviation**. Once this usage spread, it was only a matter of time until W. F. Sheppard published his famous probability table in 1903, based on the standard deviation unit, which has now become known as the table of z values.

Although use of the normal curve continued as an important tool in measuring error (deviation) from an expected population value, it was not until the 1920's, that Egon Pearson (son of Karl) formally stated the most efficient method of hypothesis testing, involving the test of one specific population value against a specific alternative population value. This method led to the definitions of **null and alternative hypotheses, type I and type II errors**, and the terms **power** and **efficiency**.

Most of modern statistics has developed since the turn of the century. With the publication of the tables of normal probabilities, statistics began to take a firm foothold in academia. By 1900 more than ten universities and colleges around the country were teaching courses in statistics, most of which were in social science departments. Within 20 years, statistics pervaded nearly every department in the social, biological, and physical sciences, in scores of schools.

It was in the year 1906, only a short time after the publication of the table of normal probabilities, that William Gossett (1876-1937) contributed one of the first, rather startling, discoveries concerning modern sampling theory. Gossett was employed by the famous Guinness Brewery in Ireland, doing chemical research and, having worked there part time during his graduate studies, he was well aware that using small samples in testing procedures led to considerable error when estimating a population mean. Drawing on his knowledge of statistics, he derived a correction formula for this estimation and in 1908 published a paper entitled "On the Probable Error of the Mean." However since there were exceedingly strict regulations on publication at Guinness, Gossett was forced to write under a pen name; he chose the name "Student."

Perhaps because he was not a professed mathematician, and because he was writing anonymously, Student's work received little attention initially. By the 1920's, however, the famous R. A. Fisher (1890-1962) had recognized this valuable contribution and was hard at work expanding this and other topics. Fisher was one of the most prolific writers in the field of statistics, and it is not unfair to say that half of all of its important concepts were developed by him. The focal point of his 100 or more articles was the publication of a book in 1925 called *Statistical Methods for Research Workers*. Not only did this work bring together many of his important contributions but it also fostered widespread diffusion of the use of statistics in research.

By the time of Fisher's work Student had derived what became known as the **t statistic** and had subsequently published tables of the *t* distribution. Fisher drew on this and his own work to contribute at least three significant ideas in his book. The first was the idea of **point estimation**. Along with Neyman and Pearson's work in the 1920's these ideas became the standard for the best test of a hypothesis for a single mean. A corollary to the idea of point estimation occurred to Fisher, and to Neyman and Pearson almost concurrently (in the 1920's). It consisted of an interval within which we could expect the population mean to exist. Fisher called this discovery the Fiducial Limits, while Neyman and Pearson called it a **Confidence Interval**. History has given the nod to Confidence Interval and that is the standard term in use today. Fisher's second contribution was to the idea of using the *t* statistic to test the difference of two means, something that had not been offered by Student. A third, and perhaps the most significant, contribution of this period was his development of the **likelihood ratio** test for the analysis of variance. Strangely enough, Fisher chose to translate this ratio into a normal score and labeled the resulting distribution z. However, in 1934 a mathematician by the name of Snededcor introduced a direct derivation of the likelihood ratio. He called this the *F distribution* after its originator, Fisher. So universally accepted was this symbol that z, which Fisher used, was discarded and later became associated with the symbol for the table of normal probabilities. These meanings for both F and z have survived today.

Nonparametric alternatives

As time passed, it became evident that tests were needed to satisfy those situations in which a normal population distribution could not be assumed or interval data were not present. A number of mathematicians worked on this problem, and although Karl Pearson proposed the celebrated **chi-square** test in 1900, most nonparametric tests were not developed until the 1940's and 1950's. In 1941, a Russian by the name of Kolmogorov published an article entitled "Confidence Limits for an Unknown Distribution" in which he derived the basic concepts now known as the Kolmogorov-Smirnov test of randomness for ordinal data. In 1945, both Mann and Wilcoxon discussed different forms of the same test in articles concerning trends in ordinal data. Wilcoxon, in several ensuing articles, extended this discussion to two sample cases in which each pair is matched. During this same period, two statisticians at the University of Chicago named Kruskal and Wallis derived a method much like Fisher's analysis of variance, which can be applied to ranked data.

The first systematic treatment of the relationship of nominal categories is also traced to Yule in 1900. In his paper, he derives what has become known as **Yule's Q**, essentially a measure of association in a cross-classification table. About the same time, in one article in 1900 and the other in 1901, Pearson developed the chi-square statistic as a goodness-of-fit test for categorical data. In 1904 he formally presented the measures of association based on chi-square (i.e., **phi** and the **contingency coefficient**) in a paper entitled "On the Correlation of Characters Not Qualitatively Measurable."

Another contributor to the problem of measurement of the relationship between attributes was the well-known R. A. Fisher (1890–1962). Among his many other contributions was the work done on cross-classification tables. In 1934 for the fourth edition of his book, *Statistics for Research Workers*, Fisher derived the exact probabilities for 2×2 tables whose cell frequencies are small (i.e., less than 5). This work was indeed useful in that this statistic yields an exact interpretation of the significance of the probability. Because of the nature of its exactness, the statistic became known as **Fisher's Exact Test**.

The first work on the relationship between ranks is credited to the psychologist Charles Spearman (1863–1945). In 1904 in an article entitled "The Proof and Measurement of Association between Two Things," he states the formula for rank-order correlation. His early work on the subject has become known as **Spearman's rank-order correlation**. Two points are worthy of note, however. First, it is apparent that Galton's initial efforts, in the development of the concept of correlation, dealt with rank ordering the size of his peas rather than using an interval scale to measure them. Therefore, it is Galton, not Spearman, who first developed the idea of rank-order correlation. Second, it was Pearson in 1907, in a paper entitled "On Further Methods of

Determining Correlation,'' who stated the present-day formula for the rank-order coefficient. It was derived by merely replacing a rank-order number for the value in an interval distribution and solving the equation. This differed significantly from Spearman's method, which used the ranks as actual values. Thus, it is Pearson, not Spearman, who is credited with the derivation. Further, since Spearman's notation for his formula directly conflicted with Yule's symbol for multiple correlation R, Pearson utilized the Greek r, rho (ρ). This symbol has survived today (although r_s is still seen in many articles).

Much has been done concerning assessment of relationships since the 1920's. Important later work on ordered data that is grouped was carried out by two University of Chicago statisticians, Goodman and Kruskal. In a series of articles in 1954, 1959, and 1963, they proposed two measures, one for frequency tables of nominal data and the other for grouped data that is ranked. The results were the statistics **gamma** and **lambda**, for ordinal and nominal data respectively. The utility of both these as measures of association is that they are easily computed and retain a very intuitive interpretation.

The field of statistics is still expanding, and no doubt there will be future R. A. Fishers and Karl Pearsons. For the interested reader who desires to know more, I recommend one book in particular: Helen Walker's *Studies in the History of Statistical Method.*

Glossary
of symbols

1. $><$ = Greater than, less than: used in conjunction with ordinal data to specify ranking, e.g., $X_1 > X_2$ (p. 12).

2. f = Frequency: the number of times a value or attribute of a particular variable occurs (p. 22).

3. i,j = Index letters: represent a counter that indicates which particular value is specified, e.g., X_i where $i = 2$ would be the second value of X (p. 22).

4. $\%$ = Percentage: indicates frequency divided by the total and multiplied by 100 (p. 23).

5. Σ = Summation: the Greek capital S, used whenever we wish to sum the values of the variable that follows it, e.g., $\Sigma_{i=1}^{n} X_i$ means we add all values of X, starting with the first and continuing until the last (p. 23).

6. cum f = Cumulative frequency: indicates accumulation of frequencies within each successive category (p. 34).

7. Mo = Mode: a measure of central tendency indicating the most frequent category or value (p. 44).

8.   Md = Median: a measure of central tendency indicating the value of the middle score or midpoint of the middle interval (p. 44).

9. $\bar{X}$ = Sample mean: defined as the sum of all values in the sample divided by N. In general it is symbolized by X or Y with the corresponding bar above the letter (p. 44).

10. μ = Population mean: the Greek letter m defined as the sum of all values in the population divided by N (p. 44).

11.   Q = Interquartile range: defined as the difference between the value at the 75th percentile and the 25th percentile (p. 54).

12.   M.D. = Mean Deviation: indicates the sum of the absolute deviations of each score from the mean, divided by N (p. 55).

13. x^2 = Squared deviation: represents $(X - \bar{X})^2$ in the formula for the variance and standard deviation. When summed across all values of X, it is referred to as the sum of the squares (p. 55).

14. σ^2 = Population variance: defined as the sum of the squared deviations in a population divided by N (p. 56).

15. σ = Population standard deviation: defined as the square root of the population variance (p. 56).

16. s^2 = Sample variance: defined as the sum of the squared deviations in the sample divided by N (p. 56).

17.   s = Sample standard deviation: defined as the square root of the sample variance. Used as an estimate of the population standard deviation (pp. 56, 125).

18. z = Standard score: indicates the number of standard deviations away from the mean a particular value is (p. 59).

19. $\mathcal{N}$ = Normal distribution: a theoretical curve which approximates certain bell-shaped frequency distributions; it is useful in computing percentiles as well as providing a foundation for inferential statistics (pp. 61, 107).

20.   Y′ = Predicted value of Y: based on the value obtained from the regression equation $Y' = a + bX$. It is the predicted value of one variable given the value of another (p. 70).

21. $a =$ Y intercept: the value on the Y axis of a graph where the regression line crosses (p. 69).

22. $b =$ Slope of the regression line: defined as the cross-product of X and Y divided by the square root of the sum of the squares of X (p. 69).

23. $s_{Y \cdot X} =$ Standard error of the estimate: indicates the standard deviation or error of prediction from the regression equation (p. 73).

24. $r^2_{XY} =$ Coefficient of determination: a measure of the amount of variance explained in one variable by knowing the corresponding values of another; it is the square of the correlation coefficient (p. 79).

25. $r_{XY} =$ Product-moment correlation: a measure of the strength and direction of the relationship between two quantitative variables (p. 80).

26. $1 - r^2_{XY} =$ Coefficient of alienation: equal to the coefficient of nondetermination or that amount of variation in one variable still unexplained, even with knowledge of a corresponding values on a second variable (p. 83).

27. $r_{XY \cdot Z} =$ Partial correlation coefficient: indicates the strength of the relationship between two variables when the effects of a third are held constant (p. 85).

28. $R_{X \cdot YZ} =$ Multiple correlation coefficient: a measure of the relationship between one variable and two others; sometimes referred to as that amount of variance in one variable explained by two others (p. 86).

29. $P(X) =$ Probability of X: defined as the proportion of favorable outcomes in a large number of trials; can also be thought of as the relative frequency of an outcome (p. 91).

30. $\infty =$ Infinity: The symbol indicating an infinitely large number. Often used in conjunction with N to indicate a limit (p. 91).

31. $! =$ Factorial: the number of orderings of N things. Defined as $N(N - 1)(N - 2)$ (p. 95).

32. $P(N, r) =$ Permutation of N things, r at a time: the number of ways to order r elements taken from a collection of N cases (p. 95).

33. $\cup =$ Union: defines the presence of one element *or* another;

in probability, it indicates the sum of probabilities as in the $P(A) + P(B)$ (p. 98).

34. $\cap$ = Intersection: defines the joint occurrence of two events; in probability, it is the product as in $P(A)$ $P(B)$ (p. 98).

35. $E(X)$ = Expected value of X: can be thought of as the average value in a sampling distribution as in the normal distribution which has an expected value of μ (p. 103).

36. $B(n,P)$ = Binomial sampling distribution: defines the theoretical distribution of two outcomes (e.g., the toss of a coin) with the probability P of occurring and N trials (p. 106).

37. $\binom{N}{r}$ = Combination of N things, r at a time: the number of ways to combine r elements from a collection of N cases (p. 106).

38. $\mathcal{N}(\mu,\sigma)$ = Normal sampling distribution: defines the theoretical distribution of values drawn independently from a population with a mean μ and standard deviation σ (p. 107).

39. $\mu_{\bar{X}}$ = Expected value of sampling distribution of means: defines the average value of sample means computed from a theoretically infinite number of samples (p. 110).

40. $\sigma_{\bar{X}}$ = Standard error of the mean: defined as the standard deviation in the sampling distribution of means and computed as the standard deviation in the population divided by the square root of N (p. 110).

41. s_X = Estimated standard error of the mean: defined as the standard deviation in the sampling distribution of means where the population variance is estimated from the unbiased sample variance; computed as S, the square root of $N - 1$ (p. 125).

42. H_0 = Null hypothesis: expressed in terms of population parameters where no systematic difference is assumed between it and the sample estimate, e.g., $\mu = 72$ (p. 113).

43. H_1 = Research or alternative hypothesis: specifies the value or direction of difference between population parameter and sample estimate, e.g., $\mu > 72$ (p. 113).

44. α = Alpha or type I error; defined as that proportion of the time a research would reject the null hypothesis when in fact it was true (p. 113).

45. β = Beta or type II error; defined as the proportion of the time a researcher would not reject the null hypothesis when in fact it was false (p. 114).

46.   z test = z ratio: it is utilized as the test statistic for the sampling distribution of means and defined as the sample mean minus the population means divided by the standard error of the mean (p. 123).

47.   d.f. = Degrees of freedom: indicates the number of values in a collection which are free to vary; specifically the size of the sample minus the number of parameters being estimated; as in $N - 1$ (p. 126).

48. $\wedge$ = Unbiased estimate: typically used in conjunction with the variance to indicate that we have computed the standard error of the mean using an unbiased sample estimate (p. 125).

49.   t test = t ratio: known as the small-sample statistic; consists of several distributions corresponding to the degrees of freedom. It is used in place of a z ratio when sample size is under 30 and population variance is unknown (p. 126).

50. $\sigma_{\bar{x}_1 - \bar{x}_2}$ = Standard error of the difference of means: defined as the standard deviation in the sampling distribution of differences of means (p. 134).

51. χ^2 = Chi square: a statistic which is distributed according to the variance of variables (pp. 138, 162).

52.   F test = F ratio: defined as the ratio of two chi-square statistics; utilized to test both the assumption of equal population variances, as well as analysis of variance problems (p. 138).

53. SS = Sum of squares: in the context of analysis of variance, indicates the squared deviation between sample means or within sample values (p. 148).

54.   MS = Mean squares: defined as the sum of squares divided by their respective degrees of freedom (p. 150).

55. $K - 1$ = Between sample degrees of freedom: K is the number of sample means and 1 is the number of parameters being estimated, i.e., the overall mean (p. 150).

56. $N - K$ = Within sample degrees of freedom: N is the number

of values in the sample and K is the number of parameters being estimated (i.e., sample means); used as the denominator for within sample mean squares (p. 151).

57. D = Smirnov single-sample test statistic: symbolizes the largest difference in cumulative proportion from a random distribution; used for nonparametric single-sample tests (p. 163).

58. K_D = Smirnov two-sample test statistic: symbolizes largest difference between two distributions of cumulative proportions (p. 165).

59. T = Wilcoxon matched-pair test statistic: defined as the sum of the ranks of the highest totaling of two ordinal variables (p. 166).

60. U = Mann-Whitney test statistic: used to measure the discrepancy between ranks for two ordinal samples (p. 167).

61. H = Kruskal-Wallis test statistic: used to test the ranking in three or more ordinal samples for significance differences (p. 168).

62. r_s = Spearman's rank-order correlation coefficient: an index of the correspondence between two ordinally ranked variables; is equivalent to Pearson's r for ranks (p. 173).

63. γ = Gamma: a measure of association for cross-classifications of ordinally ranked variables; essentially a correlation for grouped ordinal data which ignores ties (p. 175).

64. Q = Yule's measure of association: a special case of gamma for the 2×2 cross-classification table of ordinal variables (p. 176).

65. ϕ = Phi: defined as chi square divided by N. A measure of association for contingency tables based on the chi-square statistic (p. 178).

66. V = Cramer's measure of association: based on phi and often symbolized as phi prime (ϕ') (p. 179).

67. λ = Lambda: a measure of association for categorical variables; based on prediction of categories of one variable given knowledge of another (p. 180).

Algebra, the slide rule, calculators, and other forgotten topics

III

It seems true that any journey into the long-repressed subject of mathematics necessitates a revamping of our algebraic muscle. But like any muscle that has not been used for some time, it needs a workout. This appendix provides the apparatus with which to oil the mental joints and clear the cobwebs in the dusty closet of your mind.

Let's begin with the subject for which you have the longest memory lapse, the study of fractions in the fourth grade. Fractions are those pesky devils that at first glance seem relatively easy to manipulate, yet whose rules often slip by us.

1. The sum and difference of fractions:

It is true that when there is a common denominator the numerators may be combined:

$$\frac{A}{C} + \frac{B}{C} = \frac{A + B}{C} \quad \text{and} \quad \frac{A}{C} - \frac{B}{C} = \frac{A - B}{C}$$

It is definitely not true, however, that when there is a common numerator, the denominators may be combined:

$$\frac{A}{B} + \frac{A}{C} \neq \frac{A}{B + C} \quad \text{and} \quad \frac{A}{B} - \frac{A}{C} \neq \frac{A}{B + C}$$

In general, whenever we combine two fractions, the denominator is the product while the numerator is each term multiplied by the other's denominator:

$$\frac{A}{C} + \frac{B}{D} = \frac{AD + BC}{CD} \quad \text{and} \quad \frac{A}{C} - \frac{B}{D} = \frac{AD - BC}{CD}$$

2. The multiplication and division of fractions:

In general, this is carried out by multiplying straight across:

$$\frac{A}{C} \cdot \frac{B}{D} = \frac{AB}{CD}$$

Division is a little more tricky, In general, take whatever is in the denominator, invert it (i.e. switch the numerator and denominator) and multiply:

$$\frac{A}{B/C} = \frac{A}{1} \cdot \frac{C}{B} = \frac{AC}{B} \quad \text{and} \quad \frac{A/B}{C} = \frac{A}{B} \cdot \frac{1}{C} = \frac{A}{BC}$$

There is another set of rules that you may not have encountered until the tenth grade, if ever. These concern the special case of addition called summation. Summation is denoted by the Greek Σ.

1. Summing one variable:

$$\sum_{i=1}^{n} X_i = X_1 + X_2 + \ldots + X_N$$

so $$\sum_{i=2}^{5} X_i = X_2 + X_3 + X_4 + X_5$$

2. Summing combinations of variables:

$$\sum_{i=1}^{n} (X_i + Y_i) = \sum_{i=1}^{n} X_i + \sum_{i=1}^{n} Y_i$$

and $$\sum_{i=1}^{n} (X_i - Y_i) = \sum_{i=1}^{n} X_i - \sum_{i=1}^{n} Y_i$$

3. Summing powers of one variable:

$$\sum_{i=1}^{n} X_i^2 = X_1^2 + X_2^2 + \ldots + X_N^2$$

4. Summing powers of combinations of variables:

$$\sum_{i=1}^{n} (X_i + Y_i)^2 = \sum_{i=1}^{n} (X_i^2 + 2X_i Y_i + Y_i^2)$$

$$= \sum_{i=1}^{n} X_i^2 + \sum_{i=1}^{n} 2X_i Y_i + \sum_{i=1}^{n} Y_i^2 \neq \sum_{i=1}^{n} X_i^2 + \sum_{i=1}^{n} Y_i^2$$

5. Squaring sums:

$$\left(\sum_{i=1}^{n} X_i \right)^2 = (X_1 + X_2 + ... + X_N)^2 \neq \sum_{i=1}^{n} X_i^2$$

6. Summing constants:

$$\sum_{i=1}^{n} a X_i = a \cdot \sum_{i=1}^{n} X_i \quad \text{and} \quad \sum_{i=1}^{n} a_i = Na$$

7. Double summation:

$$\sum_{i=1}^{n} \sum_{j=1}^{m} X_{ij} = (X_{11} + X_{12} + ... + X_{1m}) + (X_{21} + X_{22} + ... X_{2m})$$

$$+ ... + (X_{n1} + X_{n2} + ... X_{nm})$$

A third topic that can give trouble is manipulating the special case of multiplication, that is, combining values raised to a power. For one variable, when multiplying powers, add the exponents. When dividing powers, subtract the exponents. In general:

$$X^N = X_1 \cdot X_2 \cdot X_3 ... X_n$$

1. Multiplication of powers for one variable:

$$X^A \cdot X^B = X^{A+B} \quad \text{e.g.,} \quad X^3 \cdot X^5 = X^8$$

$$X^A + X^B \neq X^{A \cdot B} \quad \text{e.g.,} \quad X^3 \cdot X^5 \neq X^{15}$$

2. Division of powers for one variable:

$$\frac{X^A}{X^B} = X^{A-B} \quad \text{e.g.,} \quad \frac{X^5}{X^3} = X^2 \quad \text{where} \quad \frac{X^A}{X^A} = X^0 = 1$$

3. Negative exponents:

$$X^{-A} = \frac{1}{X^A} \quad \text{so} \quad X^{-2} = \frac{1}{X^2}$$

4. Squaring the combination of two variables:

$$(X + Y)^2 = X^2 + 2XY + Y^2 \neq X^2 + Y^2$$

$$(X - Y)^2 = X^2 - 2XY + Y^2 \neq X^2 - Y^2$$

5. Square root of the combination of two variables:

$$\sqrt{X^2 + 2XY + Y^2} = \sqrt{(X + Y)^2} = (X + Y)$$

$$\sqrt{X^2 Y^2} = XY \quad \text{but} \quad \sqrt{X^2 + Y^2} \neq X + Y$$

No doubt all of you have had a battle with our fourth problem—finding square roots. Some of the most enlightened beginning texts include reams of pages of square root tables. I take the position that we can largely forget them. Owing to the advent and accessibility of slide rules and pocket calculators, we have very potent tools available to solve these and other problems.

The slide rule

The slide rule, costing only a few dollars (as opposed to its cousin the pocket calculator, which may cost many times that), is a quick and efficient means for solving a variety of multiplication, division, and especially square root problems. The slide rule is composed of several logarithmic scales. A logarithmic scale is one in which the product of any two numbers on a regular interval scale is equal to the sum on the log scale. The scales run from 1 to 10 and are rather spread out at the beginning, bunching up at the end:

Let's introduce this instrument by following this simple technique for multiplying. Take the equation $2 \times 3 = 6$. Find 2 on the D scale and move the slide until 1 on the C scale is opposite it. Now move the hairline to 3 on the C scale. You can read the answer on the D scale (6):

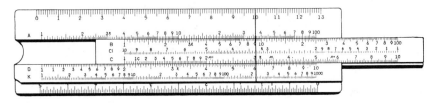

Note that this procedure would be the same if the equation read
20 × 30 = 600, or 2 × 30 = 60. The best way to tell how many places
are indicated is to round the numbers off to the nearest 10's place
and multiply to find the approximate number of places. For example,
23 × 12 = 276. We could have guessed that this would be three places
merely by multiplying 10 × 20 to get 200. Now try a simple case of
division.

To divide, we merely reverse the process of multiplication. Take
25 ÷ 5 = 5. Locate the number you are dividing *into* on the D scale;
in this case it's 25. Place opposite it on the C scale that number that
you are dividing *by*; in this case it's 5. Now read the answer on the
D scale opposite the 1 on scale C (5):

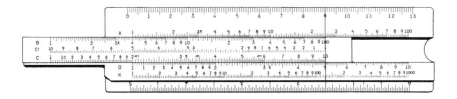

The beauty of the slide rule is that it gets easier to use for more
complicated problems. When finding the square or square root of a
number, we don't even need the center slide. In fact, it leads to less
confusion if we take it out altogether when computing these values.
To find the square of a number, merely move the hairline to the number
on the D scale and read the answer for its square on the A scale.
The square of 3 is 9. Move the hairline to 3 on the D scale and read
9 on the A scale.

The reverse is a bit trickier. Notice that the A scale has two sets
of numbers 1 to 10. How do we know which one to use in setting
the hairline on the squared value and reading the square root on the
D scale? The answer is very simple. If the number of digits (not the
value of the digits) is odd, use the first set. The number 256 has 3
digits, an odd number. Therefore, look it up using the left side of the

A scale and read the answer, 16, on the D scale. The number 64 has 2 digits, an even number, so the right side of the A scale is used to find its square root, which is 8.

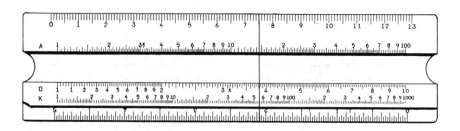

How do we know how many whole-number digits our result contains? The answer is found in one of the fundamental curiosities of arithmetic. If a squared number has one or two whole number digits, the square root will always have one digit. If the squared value has three or four digits, the square root will have two digits, and so on by twos ad infinitum.

Squared Value Number of Digits	Square Root Value Number of Digits
1 or 2	1
3 or 4	2
5 or 6	3
7 or 8	4
etc.	etc.

For example, the number 9216 has four digits and the square root is 95, two digits. The 47,961 has five digits and the square root is 219, three digits. Remember, this rule holds for whole numbers. If a decimal appears in the number, be sure to include it in the appropriate place. The number 1451.61 has four whole-number digits and so the square root must be two whole-number digits. Its value is 38.1.

There is no end to the fun in discovering the shortcut methods of the slide rule. The rule finds a use in statistics as well as in numerous other fields. Anyone interested in pursuing his knowledge of this pocket machine is directed to the paperback by Isaac Asimov entitled *An Easy Introduction to the Slide Rule,* Fawcett Books, 1965. This book is an excellent introduction to the wonders of a logarithm scale through the eyes of one of America's most popular scientists and writers.

The pocket calculator

If you are fortunate enough to own a pocket calculator, or can borrow one, you can speed these operations even more. Luckily, the technology of miniature electronics and mass production has not only reduced the size of these ingenious devices but, more important, has brought them within the price range of many students.

The principle behind a pocket calculator is really quite simple. Each time a particular number is pressed, an electronic impulse passes this information to a storage area where a certain arithmetic operation is performed. Once the answer is computed, it is displayed on the viewing screen by means of small electronic devices that emit light. Such devices, called light-emitting diodes, form the core of visual displays in calculators. In addition, some calculators, such as the one shown on this page, possess an additional memory that can store, accumulate, and recall numbers from the keyboard. For our purposes, then, the pocket calculator becomes a convenient way to add, subtract, multiply, divide, and find percentages, squares, and square roots, with accuracy to as many as 7 decimal places.

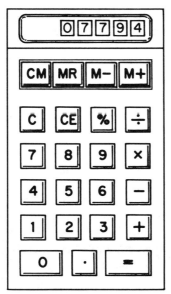

Let's try an example. In most calculators, simple addition, subtraction, multiplication, and division are carried out quite easily by entering one number, pressing the appropriate operation key (e.g., +), the second number, and finally the equals sign. That is, the number is entered in the same manner as it is represented in an equation. Thus: $15 \div 3 = 5$ becomes

	PRESS	READOUT DISPLAY
(1)	15	15
(2)	÷	15
(3)	3	3
(4)	=	5

In addition, compound operations make good use of the memory key. Consider $5 \div (1/3)$. We would perform the following:

	PRESS	READOUT DISPLAY
(1)	1	1
(2)	÷	1
(3)	3	3
(4)	=	.33333
(5)	M+	Stores .33333 in memory
(6)	5	5
(7)	÷	5
(8)	M	.33333
(9)	=	15

To find a percentage, merely multiply the number by the decimal equivalent of the percentage. For example 5% of 25 would be:

	PRESS	READOUT DISPLAY
(1)	.05	.05
(2)	X	.05
(3)	25	25
(4)	=	1.25

Of course if your calculator comes equipped with a % key, just hit 5 and then the % key before multiplying.

Finding the power or square root of a number may be a bit more complicated if no specific key such as x^y or $\sqrt{x}$ is provided. It is manageable nevertheless. Suppose you want 2 to the 4th power. You can do it on a simple calculator by multiplying $2 \times 2 \times 2 \times 2 =$. If your machine has a constant, however, merely put 2 on the display, then (with the constant on) push $\times = = = =$. For negative exponents proceed in the same manner, except push the division key instead of multiplication and push the equals key one additional time. Hence 5^{-3} would be $5 \div = = = = $.

Square roots are found by entering the number on the display and pressing the square root sign to calculate the answer. If no $\sqrt{}$ key is built in, determine the square root by the method of successive guesses. For example, suppose I want the square root of 133. Perform the following:

	PROCEDURE	KEY IN	READOUT DISPLAY
(1)	Divide 133 by 1st guess	$133 \div 11$	12.090909
(2)	Average result & 1st guess	$12.090909 + 11 \div 2 =$	11.545454
(3)	Divide 133 by answer in (2)	$133 \div 11.545454 =$	11.519685
(4)	Average the result with answer in (2)	$11.519685 + 11.545454$ $\div 2 =$	11.532569

This should provide an extremely close answer by the third guess.

(5)	To check, square it:	$11.532569 \times 11.532569 =$	133.00014

You should see immediately that these simple procedures can be combined to produce useful ways for solving various statistical computations. For example, compute the standard deviation of [7, 8, 9, 7, 12]:

	PROCEDURE	KEY IN	READOUT DISPLAY
(1)	Add values	$7 + 8 + 10 + 7 + 13$ $=$	45
(2)	Divide to find mean	$\div 5 =$	9
(3)	Subtract mean from each value, square it, and accumulate in memory	$(7 - 9)^2 =$ $M+$ $(8 - 9)^2 =$ $M+$ $(10 - 9)^2 =$ $M+$ $(7 - 9)^2 =$ $M+$ $(13 - 9)^2 =$ $M+$	4 1 1 4 16
(4)	Divide content of M by 5	$26 - 5 =$	5.2
(5)	Find square root of this variance	$5.2 \sqrt{}$	2.28035

Of course we could have computed this value by the computational formula with equal success. Either way, in much less time than it takes to do these calculations by hand, we have found the answer to our standard deviation.

For those interested in these techniques as well as some fascinating games to be played on these electronic wonders, I recommend *Games Calculators Play* by Wallace Judd, Warner Books, 1975. Meanwhile, concentrate on even more ways to use the calculator to eliminate the tedium of computation in statistics so that its intuitive purpose may be appreciated.

Tables of significance

APPENDIX *IV*

TABLE A Table of Probabilities Associated with Values as Extreme as Observed Values of z in the Normal Distribution

The body of the table gives one-tailed probabilities under H_0 of z. The left-hand marginal column gives various values of z to one decimal place. The top row gives various values to the second decimal place. Thus, for example, the one-tailed p of $z \geq .11$ or $z \leq -.11$ is $p = .4562$.

z	.00	.01	.02	.03	.04	.05	.06	.07	.08	.09
.0	.5000	.4960	.4920	.4880	.4840	.4801	.4761	.4721	.4681	.4641
.1	.4602	.4562	.4522	.4483	.4443	.4404	.4364	.4325	.4286	.4247
.2	.4207	.4168	.4129	.4090	.4052	.4013	.3974	.3936	.3897	.3859
.3	.3821	.3783	.3745	.3707	.3669	.3632	.3594	.3557	.3520	.3483
.4	.3446	.3409	.3372	.3336	.3300	.3264	.3228	.3192	.3156	.3121
.5	.3085	.3050	.3015	.2981	.2946	.2912	.2877	.2843	.2810	.2776
.6	.2743	.2709	.2676	.2643	.2611	.2578	.2546	.2514	.2483	.2451
.7	.2420	.2389	.2358	.2327	.2296	.2266	.2236	.2206	.2177	.2148
.8	.2119	.2090	.2061	.2033	.2005	.1977	.1949	.1922	.1894	.1867
.9	.1841	.1814	.1788	.1762	.1736	.1711	.1685	.1660	.1635	.1611
1.0	.1587	.1562	.1539	.1515	.1492	.1469	.1446	.1423	.1401	.1379
1.1	.1357	.1335	.1314	.1292	.1271	.1251	.1230	.1210	.1190	.1170
1.2	.1151	.1131	.1112	.1093	.1075	.1056	.1038	.1020	.1003	.0985
1.3	.0968	.0951	.0934	.0918	.0901	.0885	.0869	.0853	.0838	.0823
1.4	.0808	.0793	.0778	.0764	.0749	.0735	.0721	.0708	.0694	.0681
1.5	.0668	.0655	.0643	.0630	.0618	.0606	.0594	.0582	.0571	.0559
1.6	.0548	.0537	.0526	.0516	.0505	.0495	.0485	.0475	.0465	.0455
1.7	.0446	.0436	.0427	.0418	.0409	.0401	.0392	.0384	.0375	.0367
1.8	.0359	.0351	.0344	.0336	.0329	.0322	.0314	.0307	.0301	.0294
1.9	.0287	.0281	.0274	.0268	.0262	.0256	.0250	.0244	.0239	.0233
2.0	.0228	.0222	.0217	.0212	.0207	.0202	.0197	.0192	.0188	.0183
2.1	.0179	.0174	.0170	.0166	.0162	.0158	.0154	.0150	.0146	.0143
2.2	.0139	.0136	.0132	.0129	.0125	.0122	.0119	.0116	.0113	.0110
2.3	.0107	.0104	.0102	.0099	.0096	.0094	.0091	.0089	.0087	.0084
2.4	.0082	.0080	.0078	.0075	.0073	.0071	.0069	.0068	.0066	.0064
2.5	.0062	.0060	.0059	.0057	.0055	.0054	.0052	.0051	.0049	.0048
2.6	.0047	.0045	.0044	.0043	.0041	.0040	.0039	.0038	.0037	.0036
2.7	.0035	.0034	.0033	.0032	.0031	.0030	.0029	.0028	.0027	.0026
2.8	.0026	.0025	.0024	.0023	.0023	.0022	.0021	.0021	.0020	.0019
2.9	.0019	.0018	.0018	.0017	.0016	.0016	.0015	.0015	.0014	.0014
3.0	.0013	.0013	.0013	.0012	.0012	.0011	.0011	.0011	.0010	.0010
3.1	.0010	.0009	.0009	.0009	.0008	.0008	.0008	.0008	.0007	.0007
3.2	.0007									
3.3	.0005									
3.4	.0003									
3.5	.00023									
3.6	.00016									
3.7	.00011									
3.8	.00007									
3.9	.00005									
4.0	.00003									

From S. Siegel, *Nonparametric Statistics*, McGraw-Hill Book Company, New York, 1956, Table A with the kind permission of the publisher.

TABLE B Table of Critical Values of t^*

df	\.10	\.05	\.025	\.01	\.005	\.0005
			Level of significance for one-tailed test			
	\.20	\.10	\.05	\.02	\.01	\.001
			Level of significance for two-tailed test			
1	3.078	6.314	12.706	31.821	63.657	636.619
2	1.886	2.920	4.303	6.965	9.925	31.598
3	1.638	2.353	3.182	4.541	5.841	12.941
4	1.533	2.132	2.776	3.747	4.604	8.610
5	1.476	2.015	2.571	3.365	4.032	6.859
6	1.440	1.943	2.447	3.143	3.707	5.959
7	1.415	1.895	2.365	2.998	3.499	5.405
8	1.397	1.860	2.306	2.896	3.355	5.041
9	1.383	1.833	2.262	2.821	3.250	4.781
10	1.372	1.812	2.228	2.764	3.169	4.587
11	1.363	1.796	2.201	2.718	3.106	4.437
12	1.356	1.782	2.179	2.681	3.055	4.318
13	1.350	1.771	2.160	2.650	3.012	4.221
14	1.345	1.761	2.145	2.624	2.977	4.140
15	1.341	1.753	2.131	2.602	2.947	4.073
16	1.337	1.746	2.120	2.583	2.921	4.015
17	1.333	1.740	2.110	2.567	2.898	3.965
18	1.330	1.734	2.101	2.552	2.878	3.922
19	1.328	1.729	2.093	2.539	2.861	3.883
20	1.325	1.725	2.086	2.528	2.845	3.850
21	1.323	1.721	2.080	2.518	2.831	3.819
22	1.321	1.717	2.074	2.508	2.819	3.792
23	1.319	1.714	2.069	2.500	2.807	3.767
24	1.318	1.711	2.064	2.492	2.797	3.745
25	1.316	1.708	2.060	2.485	2.787	3.725
26	1.315	1.706	2.056	2.479	2.779	3.707
27	1.314	1.703	2.052	2.473	2.771	3.690
28	1.313	1.701	2.048	2.467	2.763	3.674
29	1.311	1.699	2.045	2.462	2.756	3.659
30	1.310	1.697	2.042	2.457	2.750	3.646
40	1.303	1.684	2.021	2.423	2.704	3.551
60	1.296	1.671	2.000	2.390	2.660	3.460
120	1.289	1.658	1.980	2.358	2.617	3.373
∞	1.282	1.645	1.960	2.326	2.576	3.291

Abridged from Table III of Fisher and Yates, *Statistical Tables for Biological, Agricultural, and Medical Research*, published by Oliver and Boyd Ltd., Edinburgh, by permission of the authors and publishers.

TABLE C Table of Critical Values of Chi Square*

df	.99	.98	.95	.90	.80	.70	.50	.30	.20	.10	.05	.02	.01	.001
				Probability under H_0 that $\chi^2 \geq$ chi square									
1	.00016	.00063	.0039	.016	.064	.15	.46	1.07	1.64	2.71	3.84	5.41	6.64	10.83
2	.02	.04	.10	.21	.45	.71	1.39	2.41	3.22	4.60	5.99	7.82	9.21	13.82
3	.12	.18	.35	.58	1.00	1.42	2.37	3.66	4.64	6.25	7.82	9.84	11.34	16.27
4	.30	.43	.71	1.06	1.65	2.20	3.36	4.88	5.99	7.78	9.49	11.67	13.28	18.46
5	.55	.75	1.14	1.61	2.34	3.00	4.35	6.06	7.29	9.24	11.07	13.39	15.09	20.52
6	.87	1.13	1.64	2.20	3.07	3.83	5.35	7.23	8.56	10.64	12.59	15.03	16.81	22.46
7	1.24	1.56	2.17	2.83	3.82	4.67	6.35	8.38	9.80	12.02	14.07	16.62	18.48	24.32
8	1.65	2.03	2.73	3.49	4.59	5.53	7.34	9.52	11.03	13.36	15.51	18.17	20.09	26.12
9	2.09	2.53	3.32	4.17	5.38	6.39	8.34	10.66	12.24	14.68	16.92	19.68	21.67	27.88
10	2.56	3.06	3.94	4.86	6.18	7.27	9.34	11.78	13.44	15.99	18.31	21.16	23.21	29.59
11	3.05	3.61	4.58	5.58	6.99	8.15	10.34	12.90	14.63	17.28	19.68	22.62	24.72	31.26
12	3.57	4.18	5.23	6.30	7.81	9.03	11.34	14.01	15.81	18.55	21.03	24.05	26.22	32.91
13	4.11	4.76	5.89	7.04	8.63	9.93	12.34	15.12	16.98	19.81	22.36	25.47	27.69	34.53
14	4.66	5.37	6.57	7.79	9.47	10.82	13.34	16.22	18.15	21.06	23.68	26.87	29.14	36.12
15	5.23	5.98	7.26	8.55	10.31	11.72	14.34	17.32	19.31	22.31	25.00	28.26	30.58	37.70
16	5.81	6.61	7.96	9.31	11.15	12.62	15.34	18.42	20.46	23.54	26.30	29.63	32.00	39.29
17	6.41	7.26	8.67	10.08	12.00	13.53	16.34	19.51	21.62	24.77	27.59	31.00	33.41	40.75
18	7.02	7.91	9.39	10.86	12.86	14.44	17.34	20.60	22.76	25.99	28.87	32.35	34.80	42.31
19	7.63	8.57	10.12	11.65	13.72	15.35	18.34	21.69	23.90	27.20	30.14	33.69	36.19	43.82
20	8.26	9.24	10.85	12.44	14.58	16.27	19.34	22.78	25.04	28.41	31.41	35.02	37.57	45.32
21	8.90	9.92	11.59	13.24	15.44	17.18	20.34	23.86	26.17	29.62	32.67	36.34	38.93	46.80
22	9.54	10.60	12.34	14.04	16.31	18.10	21.24	24.94	27.30	30.81	33.92	37.66	40.29	48.27
23	10.20	11.29	13.09	14.85	17.19	19.02	22.34	26.02	28.43	32.01	35.17	38.97	41.64	49.73
24	10.86	11.99	13.85	15.66	18.06	19.94	23.34	27.10	29.55	33.20	36.42	40.27	42.98	51.18
25	11.52	12.70	14.61	16.47	18.94	20.87	24.34	28.17	30.68	34.38	37.65	41.57	44.31	52.62
26	12.20	13.41	15.38	17.29	19.82	21.79	25.34	29.25	31.80	35.56	38.88	42.86	45.64	54.05
27	12.88	14.12	16.15	18.11	20.70	22.72	26.34	30.32	32.91	36.74	40.11	44.14	46.96	55.48
28	13.56	14.85	16.93	18.94	21.59	23.65	27.34	31.39	34.03	37.92	41.34	45.42	48.28	56.89
29	14.26	15.57	17.71	19.77	22.48	24.58	28.34	32.46	35.14	39.09	42.56	46.69	49.59	58.30
30	14.95	16.31	18.49	20.60	23.36	25.51	29.34	33.53	36.25	40.26	43.77	47.96	50.89	59.70

Abridged from Table IV of Fisher and Yates, *Statistical tables for biological, agricultural, and medical research,* published by Oliver and Boyd Ltd., Edinburgh, by permission of the authors and publishers.

TABLE D Distribution of F^*

$$p = .05$$

n_1 / n_2	1	2	3	4	5	6	8	12	24	∞
1	161.4	199.5	215.7	224.6	230.2	234.0	238.9	243.9	249.0	254.3
2	18.51	19.00	19.16	19.25	19.30	19.33	19.37	19.41	19.45	19.50
3	10.13	9.55	9.28	9.12	9.01	8.94	8.84	8.74	8.64	8.53
4	7.71	6.94	6.59	6.39	6.26	6.16	6.04	5.91	5.77	5.63
5	6.61	5.79	5.41	5.19	5.05	4.95	4.82	4.68	4.53	4.36
6	5.99	5.14	4.76	4.53	4.39	4.28	4.15	4.00	3.84	3.67
7	5.59	4.74	4.35	4.12	3.97	3.87	3.73	3.57	3.41	3.23
8	5.32	4.46	4.07	3.84	3.69	3.58	3.44	3.28	3.12	2.93
9	5.12	4.26	3.86	3.63	3.48	3.37	3.23	3.07	2.90	2.71
10	4.96	4.10	3.71	3.48	3.33	3.22	3.07	2.91	2.74	2.54
11	4.84	3.98	3.59	3.36	3.20	3.09	2.95	2.79	2.61	2.40
12	4.75	3.88	3.49	3.26	3.11	3.00	2.85	2.69	2.50	2.30
13	4.67	3.80	3.41	3.18	3.02	2.92	2.77	2.60	2.42	2.21
14	4.60	3.74	3.34	3.11	2.96	2.85	2.70	2.53	2.35	2.13
15	4.54	3.68	3.29	3.06	2.90	2.79	2.64	2.48	2.29	2.07
16	4.49	3.63	3.24	3.01	2.85	2.74	2.59	2.42	2.24	2.01
17	4.45	3.59	3.20	2.96	2.81	2.70	2.55	2.38	2.19	1.96
18	4.41	3.55	3.16	2.93	2.77	2.66	2.51	2.34	2.15	1.92
19	4.38	3.52	3.13	2.90	2.74	2.63	2.48	2.31	2.11	1.88
20	4.35	3.49	3.10	2.87	2.71	2.60	2.45	2.28	2.08	1.84
21	4.32	3.47	3.07	2.84	2.68	2.57	2.42	2.25	2.05	1.81
22	4.30	3.44	3.05	2.82	2.66	2.55	2.40	2.23	2.03	1.78
23	4.28	3.42	3.03	2.80	2.64	2.53	2.38	2.20	2.00	1.76
24	4.26	3.40	3.01	2.78	2.62	2.51	2.36	2.18	1.98	1.73
25	4.24	3.38	2.99	2.76	2.60	2.49	2.34	2.16	1.96	1.71
26	4.22	3.37	2.98	2.74	2.59	2.47	2.32	2.15	1.95	1.69
27	4.21	3.35	2.96	2.73	2.57	2.46	2.30	2.13	1.93	1.67
28	4.20	3.34	2.95	2.71	2.56	2.44	2.29	2.12	1.91	1.65
29	4.18	3.33	2.93	2.70	2.54	2.43	2.28	2.10	1.90	1.64
30	4.17	3.32	2.92	2.69	2.53	2.42	2.27	2.09	1.89	1.62
40	4.08	3.23	2.84	2.61	2.45	2.34	2.18	2.00	1.79	1.51
60	4.00	3.15	2.76	2.52	2.37	2.25	2.10	1.92	1.70	1.39
120	3.92	3.07	2.68	2.45	2.29	2.17	2.02	1.83	1.61	1.25
∞	3.84	2.99	2.60	2.37	2.21	2.09	1.94	1.75	1.52	1.00

Values of n_1 and n_2 represent the degrees of freedom associated with the larger and smaller estimates of variance respectively.

*Abridged from Table V of R. A. Fisher and F. Yates, *Statistical Tables for Biological, Agricultural and Medical Research* (1948 ed.), published by Oliver & Boyd, Ltd., Edinburgh and London, by permission of the authors and publishers.

TABLE D   Distribution of *F* (*Continued*)

$$p = .01$$

n_2 \ n_1	1	2	3	4	5	6	8	12	24	∞
1	4052	4999	5403	5625	5764	5859	5981	6106	6234	6366
2	98.49	99.01	99.17	99.25	99.30	99.33	99.36	99.42	99.46	99.50
3	34.12	30.81	29.46	28.71	28.24	27.91	27.49	27.05	26.60	26.12
4	21.20	18.00	16.69	15.98	15.52	15.21	14.80	14.37	13.93	13.46
5	16.26	13.27	12.06	11.39	10.97	10.67	10.27	9.89	9.47	9.02
6	13.74	10.92	9.78	9.15	8.75	8.47	8.10	7.72	7.31	6.88
7	12.25	9.55	8.45	7.85	7.46	7.19	6.84	6.47	6.07	5.65
8	11.26	8.65	7.59	7.01	6.63	6.37	6.03	5.67	5.28	4.86
9	10.56	8.02	6.99	6.42	6.06	5.80	5.47	5.11	4.73	4.31
10	10.04	7.56	6.55	5.99	5.64	5.39	5.06	4.71	4.33	3.91
11	9.65	7.20	6.22	5.67	5.32	5.07	4.74	4.40	4.02	3.60
12	9.33	6.93	5.95	5.41	5.06	4.82	4.50	4.16	3.78	3.36
13	9.07	6.70	5.74	5.20	4.86	4.62	4.30	3.96	3.59	3.16
14	8.86	6.51	5.56	5.03	4.69	4.46	4.14	3.80	3.43	3.00
15	8.68	6.36	5.42	4.89	4.56	4.32	4.00	3.67	3.29	2.87
16	8.53	6.23	5.29	4.77	4.44	4.20	3.89	3.55	3.18	2.75
17	8.40	6.11	5.18	4.67	4.34	4.10	3.79	3.45	3.08	2.65
18	8.28	6.01	5.09	4.58	4.25	4.01	3.71	3.37	3.00	2.57
19	8.18	5.93	5.01	4.50	4.17	3.94	3.63	3.30	2.92	2.49
20	8.10	5.85	4.94	4.43	4.10	3.87	3.56	3.23	2.86	2.42
21	8.02	5.78	4.87	4.37	4.04	3.81	3.51	3.17	2.80	2.36
22	7.94	5.72	4.82	4.31	3.99	3.76	3.45	3.12	2.75	2.31
23	7.88	5.66	4.76	4.26	3.94	3.71	3.41	3.07	2.70	2.26
24	7.82	5.61	4.72	4.22	3.90	3.67	3.36	3.03	2.66	2.21
25	7.77	5.57	4.68	4.18	3.86	3.63	3.32	2.99	2.62	2.17
26	7.72	5.53	4.64	4.14	3.82	3.59	3.29	2.96	2.58	2.13
27	7.68	5.49	4.60	4.11	3.78	3.56	3.26	2.93	2.55	2.10
28	7.64	5.45	4.57	4.07	3.75	3.53	3.23	2.90	2.52	2.06
29	7.60	5.42	4.54	4.04	3.73	3.50	3.20	2.87	2.49	2.03
30	7.56	5.39	4.51	4.02	3.70	3.47	3.17	2.84	2.47	2.01
40	7.31	5.18	4.31	3.83	3.51	3.29	2.99	2.66	2.29	1.80
60	7.08	4.98	4.13	3.65	3.34	3.12	2.82	2.50	2.12	1.60
120	6.85	4.79	3.95	3.48	3.17	2.96	2.66	2.34	1.95	1.38
∞	6.64	4.60	3.78	3.32	3.02	2.80	2.51	2.18	1.79	1.00

Values of n_1 and n_2 represent the degrees of freedom associated with the larger and smaller estimates of variance respectively.

TABLE E   Table of Critical Values of D in the Kolmogorov-Smirnov One-sample Test*

Sample size (N)	Level of significance for $D = \text{maximum} \ |F_0(X) - S_N(X)|$				
	.20	.15	.10	.05	.01
1	.900	.925	.950	.975	.995
2	.684	.726	.776	.842	.929
3	.565	.597	.642	.708	.828
4	.494	.525	.564	.624	.733
5	.446	.474	.510	.565	.669
6	.410	.436	.470	.521	.618
7	.381	.405	.438	.486	.577
8	.358	.381	.411	.457	.54˜
9	.339	.360	.388	.432	.51˙.
10	.322	.342	.368	.410	.490
11	.307	.326	.352	.391	.468
12	.295	.313	.338	.375	.450
13	.284	.302	.325	.361	.433
14	.274	.292	.314	.349	.418
15	.266	.283	.304	.338	.404
16	.258	.274	.295	.328	.392
17	.250	.266	.286	.318	.381
18	.244	.259	.278	.309	.371
19	.237	.252	.272	.301	.363
20	.231	.246	.264	.294	.356
25	.21	.22	.24	.27	.32
30	.19	.20	.22	.24	.29
35	.18	.19	.21	.23	.27
Over 35	$\dfrac{1.07}{\sqrt{N}}$	$\dfrac{1.14}{\sqrt{N}}$	$\dfrac{1.22}{\sqrt{N}}$	$\dfrac{1.36}{\sqrt{N}}$	$\dfrac{1.63}{\sqrt{N}}$

*Adapted from Massey, F. J., Jr. 1951. The Kolmogorov-Smirnov test for goodness of fit. *J. Amer. Statist. Ass.*, **46**, 70, with the kind permission of the author and publisher.

TABLE F Table of Critical Values of K_D in the Kolmogorov-Smirnov Two-sample Test

(Small samples)

N	One-tailed test*		Two-tailed test†	
	$\alpha = .05$	$\alpha = .01$	$\alpha = .05$	$\alpha = .01$
3	3	—	—	—
4	4	—	4	—
5	4	5	5	5
6	5	6	5	6
7	5	6	6	6
8	5	6	6	7
9	6	7	6	7
10	6	7	7	8
11	6	8	7	8
12	6	8	7	8
13	7	8	7	9
14	7	8	8	9
15	7	9	8	9
16	7	9	8	10
17	8	9	8	10
18	8	10	9	10
19	8	10	9	10
20	8	10	9	11
21	8	10	9	11
22	9	11	9	11
23	9	11	10	11
24	9	11	10	12
25	9	11	10	12
26	9	11	10	12
27	9	12	10	12
28	10	12	11	13
29	10	12	11	13
30	10	12	11	13
35	11	13	12	
40	11	14	13	

*Abridged from Goodman, L. A. 1954. Kolmogorov-Smirnov tests for psychological research. *Psychol. Bull.*, **51**, 167, with the kind permission of the author and the American Psychological Association.

†Derived from Table 1 of Massey, F. J., Jr. 1951. The distribution of the maximum deviation between two sample cumulative step functions. *Ann. Math. Statist.*, **22**, 126-127, with the kind permission of the author and the publisher.

TABLE G₁ Critical Values of U and U' for a One-tailed Test a $\alpha = 0.025$ or a Two-tailed Test at $\alpha = 0.05$

To be significant for any given n_1 and n_2: Obtained U must be equal to or <u>less than</u> the value shown in the table. Obtained U' must be equal to or <u>greater than</u> the value shown in the table.

Each cell shows the value of U (upper) and U' (lower).

n_2 \ n_1	1	2	3	4	5	6	7	8	9	10	11	12	13	14	15	16	17	18	19	20
1	--	--	--	--	--	--	--	--	--	--	--	--	--	--	--	--	--	--	--	--
2	--	--	--	--	--	--	--	0/16	0/18	0/20	0/22	1/23	1/25	1/27	1/29	1/31	2/32	2/34	2/36	2/38
3	--	--	--	--	0/15	1/17	1/20	2/22	2/25	3/27	3/30	4/32	4/35	5/37	5/40	6/42	6/45	7/47	7/50	8/52
4	--	--	--	0/16	1/19	2/22	3/25	4/28	4/32	5/35	6/38	7/41	8/44	9/47	10/50	11/53	11/57	12/60	13/63	13/67
5	--	--	0/15	1/19	2/23	3/27	5/30	6/34	7/38	8/42	9/46	11/49	12/53	13/57	14/61	15/65	17/68	18/72	19/76	20/80
6	--	--	1/17	2/22	3/27	5/31	6/36	8/40	10/44	11/49	13/53	14/58	16/62	17/67	19/71	21/75	22/80	24/84	25/89	27/93
7	--	--	1/20	3/25	5/30	6/36	8/41	10/46	12/51	14/56	16/61	18/66	20/71	22/76	24/81	26/86	28/91	30/96	32/101	34/106
8	--	0/16	2/22	4/28	6/34	8/40	10/46	13/51	15/57	17/63	19/69	22/74	24/80	26/86	29/91	31/97	34/102	36/108	38/111	41/119
9	--	0/18	2/25	4/32	7/38	10/44	12/51	15/57	17/64	20/70	23/76	26/82	28/89	31/95	34/101	37/107	39/114	42/120	45/126	48/132
10	--	0/20	3/27	5/35	8/42	11/49	14/56	17/63	20/70	23/77	26/84	29/91	33/97	36/104	39/111	42/118	45/125	48/132	52/138	55/145
11	--	0/22	3/30	6/38	9/46	13/53	16/61	19/69	23/76	26/84	30/91	33/99	37/106	40/114	44/121	47/129	51/136	55/143	58/151	62/158
12	--	1/23	4/32	7/41	11/49	14/58	18/66	22/74	26/82	29/91	33/99	37/107	41/115	45/123	49/131	53/139	57/147	61/155	65/163	69/171
13	--	1/25	4/35	8/44	12/53	16/62	20/71	24/80	28/89	33/97	37/106	41/115	45/124	50/132	54/141	59/149	63/158	67/167	72/175	76/184
14	--	1/27	5/37	9/47	13/51	17/67	22/76	26/86	31/95	36/104	40/114	45/123	50/132	55/141	59/151	64/160	67/171	74/178	78/188	83/197
15	--	1/29	5/40	10/50	14/61	19/71	24/81	29/91	34/101	39/111	44/121	49/131	54/141	59/151	64/161	70/170	75/180	80/190	85/200	90/210
16	--	1/31	6/42	11/53	15/65	21/75	26/86	31/97	37/107	42/118	47/129	53/139	59/149	64/160	70/170	75/181	81/191	86/202	92/212	98/222
17	--	2/32	6/45	11/57	17/68	22/80	28/91	34/102	39/114	45/125	51/136	57/147	63/158	67/171	75/180	81/191	87/202	93/213	99/224	105/235
18	--	2/34	7/47	12/60	18/72	24/84	30/96	36/108	42/120	48/132	55/143	61/155	67/167	74/178	80/190	86/202	93/213	99/225	106/236	112/248
19	--	2/36	7/50	13/63	19/76	25/89	32/101	38/114	45/126	52/138	58/151	65/163	72/175	78/188	85/200	92/212	99/224	106/236	113/248	119/261
20	--	2/38	8/52	13/67	20/80	27/93	34/106	41/119	48/132	55/145	62/158	69/171	76/184	83/197	90/210	98/222	105/235	112/248	119/261	127/273

(Dashes in the body of the table indicate that no decision is possible at the stated level of significance.)

From R. Runyon and A. Haber, *Fundamentals of Behavioral Statistics*, 3rd ed. Reading, Mass.: Addison-Wesley Publishing Co., 1976, Table I, with the kind permission of the publisher.

TABLE G₂ Critical Values of U and U' for a One-tailed Test a $\alpha = 0.05$ or a Two-tailed Test at $\alpha = 0.10$

To be significant for any given n_1 and n_2: Obtained U must be equal to or less than the value shown in the table. Obtained U' must be equal to or greater than the value shown in the table.

Each cell shows the U value (upper) over the U' value (lower).

n_2 \ n_1	1	2	3	4	5	6	7	8	9	10	11	12	13	14	15	16	17	18	19	20
1	--	--	--	--	--	--	--	--	--	--	--	--	--	--	--	--	--	--	0 / 19	0 / 20
2	--	--	--	--	0 / 10	0 / 12	0 / 14	1 / 15	1 / 17	1 / 19	1 / 21	2 / 22	2 / 24	2 / 26	3 / 27	3 / 29	3 / 31	4 / 32	4 / 34	4 / 36
3	--	--	0 / 9	0 / 12	1 / 14	2 / 16	2 / 19	3 / 21	3 / 24	4 / 26	5 / 28	5 / 31	6 / 33	7 / 35	7 / 38	8 / 40	9 / 42	9 / 45	10 / 47	11 / 49
4	--	--	0 / 12	1 / 15	2 / 18	3 / 21	4 / 24	5 / 27	6 / 30	7 / 33	8 / 36	9 / 39	10 / 42	11 / 45	12 / 48	14 / 50	15 / 53	16 / 56	17 / 59	18 / 62
5	--	0 / 10	1 / 14	2 / 18	4 / 21	5 / 25	6 / 29	8 / 32	9 / 36	11 / 39	12 / 43	13 / 47	15 / 50	16 / 54	18 / 57	19 / 61	20 / 65	22 / 68	23 / 72	25 / 75
6	--	0 / 12	2 / 16	3 / 21	5 / 25	7 / 29	8 / 34	10 / 38	12 / 42	14 / 46	16 / 50	17 / 55	19 / 59	21 / 63	23 / 67	25 / 71	26 / 76	28 / 80	30 / 84	32 / 88
7	--	0 / 14	2 / 19	4 / 24	6 / 29	8 / 34	11 / 38	13 / 43	15 / 48	17 / 53	19 / 58	21 / 63	24 / 67	26 / 72	28 / 77	30 / 82	33 / 86	35 / 91	37 / 96	39 / 101
8	--	1 / 15	3 / 21	5 / 27	8 / 32	10 / 38	13 / 43	15 / 49	18 / 54	20 / 60	23 / 65	26 / 70	28 / 76	31 / 81	33 / 87	36 / 92	39 / 97	41 / 103	44 / 108	47 / 113
9	--	1 / 17	3 / 24	6 / 30	9 / 36	12 / 42	15 / 48	18 / 54	21 / 60	24 / 66	27 / 72	30 / 78	33 / 84	36 / 90	39 / 96	42 / 102	45 / 108	48 / 114	51 / 120	54 / 126
10	--	1 / 19	4 / 26	7 / 33	11 / 39	14 / 46	17 / 53	20 / 60	24 / 66	27 / 73	31 / 79	34 / 86	37 / 93	41 / 99	44 / 106	48 / 112	51 / 119	55 / 125	58 / 132	62 / 138
11	--	1 / 21	5 / 28	8 / 36	12 / 43	16 / 50	19 / 58	23 / 65	27 / 72	31 / 79	34 / 87	38 / 94	42 / 101	46 / 108	50 / 115	54 / 122	57 / 130	61 / 137	65 / 144	69 / 151
12	--	2 / 22	5 / 31	9 / 39	13 / 47	17 / 55	21 / 63	26 / 70	30 / 78	34 / 86	38 / 94	42 / 102	47 / 109	51 / 117	55 / 125	60 / 132	64 / 140	68 / 148	72 / 156	77 / 163
13	--	2 / 24	6 / 33	10 / 42	15 / 50	19 / 59	24 / 67	28 / 76	33 / 84	37 / 93	42 / 101	47 / 109	51 / 118	56 / 126	61 / 134	65 / 143	70 / 151	75 / 159	80 / 167	84 / 176
14	--	2 / 26	7 / 35	11 / 45	16 / 54	21 / 63	26 / 72	31 / 81	36 / 90	41 / 99	46 / 108	51 / 117	56 / 126	61 / 135	66 / 144	71 / 153	77 / 161	82 / 170	87 / 179	92 / 188
15	--	3 / 27	7 / 38	12 / 48	18 / 57	23 / 67	28 / 77	33 / 87	39 / 96	44 / 106	50 / 115	55 / 125	61 / 134	66 / 144	72 / 153	77 / 163	83 / 172	88 / 182	94 / 191	100 / 200
16	--	3 / 29	8 / 40	14 / 50	19 / 61	25 / 71	30 / 82	36 / 92	42 / 102	48 / 112	54 / 122	60 / 132	65 / 143	71 / 153	77 / 163	83 / 173	89 / 183	95 / 193	101 / 203	107 / 213
17	--	3 / 31	9 / 42	15 / 53	20 / 65	26 / 76	33 / 86	39 / 97	45 / 108	51 / 119	57 / 130	64 / 140	70 / 151	77 / 161	83 / 172	89 / 183	96 / 193	102 / 204	109 / 214	115 / 225
18	--	4 / 32	9 / 45	16 / 56	22 / 68	28 / 80	35 / 91	41 / 103	48 / 114	55 / 123	61 / 137	68 / 148	75 / 159	82 / 170	88 / 182	95 / 193	102 / 204	109 / 215	116 / 226	123 / 237
19	0 / 19	4 / 34	10 / 47	17 / 59	23 / 72	30 / 84	37 / 96	44 / 108	51 / 120	58 / 132	65 / 144	72 / 156	80 / 167	87 / 179	94 / 191	101 / 203	109 / 214	116 / 226	123 / 238	130 / 250
20	0 / 20	4 / 36	11 / 49	18 / 62	25 / 75	32 / 88	39 / 101	47 / 113	54 / 126	62 / 138	69 / 151	77 / 163	84 / 176	92 / 188	100 / 200	107 / 213	115 / 225	123 / 237	130 / 250	138 / 262

(Dashes in the body of the table indicate that no decision is possible at the stated level of significance.)

From R. Runyon and A. Haber, *Fundamentals of Behavioral Statistics*, 2nd ed., Addison-Wesley Publishing Company, Reading, Mass., 1971, Table I, with the kind permission of the publisher.

TABLE H Table of Critical Values of T in the Wilcoxon Matched-pairs Signed-ranks Test*

N	Level of significance for one-tailed test		
	.025	.01	.005
	Level of significance for two-tailed test		
	.05	.02	.01
6	0	—	—
7	2	0	—
8	4	2	0
9	6	3	2
10	8	5	3
11	11	7	5
12	14	10	7
13	17	13	10
14	21	16	13
15	25	20	16
16	30	24	20
17	35	28	23
18	40	33	28
19	46	38	32
20	52	43	38
21	59	49	43
22	66	56	49
23	73	62	55
24	81	69	61
25	89	77	68

*Adapted from Table I of Wilcoxon, F. 1949. *Some rapid approximate statistical procedures.* New York: American Cyanamid Company, p. 13, and with the kind permission of the author and publisher.

TABLE I Table of Probabilities Associated with Values as Large as Observed Values of H in the Kruskal-Wallis One-way Analysis of Variance by Ranks*

n_1	n_2	n_3	H	p	n_1	n_2	n_3	H	p
2	1	1	2.7000	.500	4	3	2	6.4444	.008
								6.3000	.011
2	2	1	3.6000	.200				5.4444	.046
								5.4000	.051
2	2	2	4.5714	.067				4.5111	.098
			3.7143	.200				4.4444	.102
3	1	1	3.2000	.300	4	3	3	6.7455	.010
								6.7091	.013
3	2	1	4.2857	.100				5.7909	.046
			3.8571	.133				5.7273	.050
3	2	2	5.3572	.029				4.7091	.092
			4.7143	.048				4.7000	.101
			4.5000	.067					
			4.4643	.105	4	4	1	6.6667	.010
								6.1667	.022
3	3	1	5.1429	.043				4.9667	.048
			4.5714	.100				4.8667	.054
			4.0000	.129				4.1667	.082
								4.0667	.102
3	3	2	6.2500	.011					
			5.3611	.032	4	4	2	7.0364	.006
			5.1389	.061				6.8727	.011
			4.5556	.100				5.4545	.046
			4.2500	.121				5.2364	.052
								4.5545	.098
3	3	3	7.2000	.004				4.4455	.103
			6.4889	.011					
			5.6889	.029	4	4	3	7.1439	.010
			5.6000	.050				7.1364	.011
			5.0667	.086				5.5985	.049
			4.6222	.100				5.5758	.051
								4.5455	.099
4	1	1	3.5714	.200				4.4773	.102
4	2	1	4.8214	.057					
			4.5000	.076	4	4	4	7.6538	.008
			4.0179	.114				7.5385	.011
								5.6923	.049
4	2	2	6.0000	.014				5.6538	.054
			5.3333	.033				4.6539	.097
			5.1250	.052				4.5001	.104
			4.4583	.100					
			4.1667	.105	5	1	1	3.8571	.143
4	3	1	5.8333	.021	5	2	1	5.2500	.036
			5.2083	.050				5.0000	.048
			5.0000	.057				4.4500	.071
			4.0556	.093				4.2000	.095
			3.8889	.129				4.0500	.119

TABLE I   Table of Probabilities Associated with Values as Large as Observed Values of H in the Kruskal-Wallis One-way Analysis of Variance by Ranks* (*Continued*)

n_1	n_2	n_3	H	p	n_1	n_2	n_3	H	p
5	2	2	6.5333	.008				5.6308	.050
			6.1333	.013				4.5487	.099
			5.1600	.034				4.5231	.103
			5.0400	.056	5	4	4	7.7604	.009
			4.3733	.090				7.7440	.011
			4.2933	.122				5.6571	.049
5	3	1	6.4000	.012				5.6176	.050
			4.9600	.048				4.6187	.100
			4.8711	.052				4.5527	.102
			4.0178	.095	5	5	1	7.3091	.009
			3.8400	.123				6.8364	.011
5	3	2	6.9091	.009				5.1273	.046
			6.8218	.010				4.9091	.053
			5.2509	.049				4.1091	.086
			5.1055	.052				4.0364	.105
			4.6509	.091	5	5	2	7.3385	.010
			4.4945	.101				7.2692	.010
5	3	3	7.0788	.009				5.3385	.047
			6.9818	.011				5.2462	.051
			5.6485	.049				4.6231	.097
			5.5152	.051				4.5077	.100
			4.5333	.097	5	5	3	7.5780	.010
			4.4121	.109				7.5429	.010
5	4	1	6.9545	.008				5.7055	.046
			6.8400	.011				5.6264	.051
			4.9855	.044				4.5451	.100
			4.8600	.056				4.5363	.102
			3.9873	.098	5	5	4	7.8229	.010
			3.9600	.102				7.7914	.010
5	4	2	7.2045	.009				5.6657	.049
			7.1182	.010				5.6429	.050
			5.2727	.049				4.5229	.099
			5.2682	.050				4.5200	.101
			4.5409	.098	5	5	5	8.0000	.009
			4.5182	.101				7.9800	.010
5	4	3	7.4449	.010				5.7800	.049
			7.3949	.011				5.6600	.051
			5.6564	.049				4.5600	.100
								4.5000	.102

*Adapted and abridged from Kruskal, W. H., and Wallis, W. A. 1952. Use of ranks in one-criterion variance analysis. *J. Amer. Statist. Ass.*, **47**, 614–617, with the kind permission of the authors and the publisher. (The corrections to this table given by the authors in Errata, *J. Amer. Statist. Ass.*, **48**, 910, have been incorporated.)

Index